Audio Metering

D1217958

Audio Metering
Measurements, Standards and Practice

Eddy B. Brixen

ELSEVIER

AMSTERDAM • BOSTON • HEIDELBERG • LONDON • NEW YORK • OXFORD
PARIS • SAN DIEGO • SAN FRANCISCO • SINGAPORE • SYDNEY • TOKYO

Focal Press is an Imprint of Elsevier

Focal
Press

Focal Press is an imprint of Elsevier
The Boulevard, Langford Lane, Kidlington, Oxford, OX5 1GB, UK
30 Corporate Drive, Suite 400, Burlington, MA 01803, USA

First published 2011

Notices
Knowledge and best practice in this field are constantly changing. As new research and experience
broaden our understanding, changes in research methods, professional practices, or medical
treatment may become necessary.

Practitioners and researchers must always rely on their own experience and knowledge in evaluating
and using any information, methods, compounds, or experiments described herein. In using such
information or methods they should be mindful of their own safety and the safety of others,
including parties for whom they have a professional responsibility.

To the fullest extent of the law, neither the Publisher nor the authors, contributors, or editors, assume
any liability for any injury and/or damage to persons or property as a matter of products liability,
negligence or otherwise, or from any use or operation of any methods, products, instructions, or
ideas contained in the material herein.

British Library Cataloguing in Publication Data
A catalogue record for this book is available from the British Library

Library of Congress Control Number: 2010938289

ISBN: 978-0-240-81467-4

For information on all Focal Press publications
visit our website at focalpress.com

Printed and bound in the United States

10 11 12 13 10 9 8 7 6 5 4 3 2 1

Table of Contents

Preface

What is dynamic range — and how loud is it? These are the eternal questions that concern everyone who works with the practical aspects of sound.

This book was written to give everybody with an interest in audio an explanation of the conditions that determine the answers to these questions.

Fundamental acoustics, electronics and psycho-acoustic concepts are described here. A number of topics related to digital technology are also covered and information can also be obtained here on the majority of the tools that are used in describing the magnitude of sound.

This is the second edition of *Audio Metering*. The update is particularly concerned with loudness measures and metering. However, the basic chapters in the beginning of the book have been expanded and a complete new chapter on room acoustics has been added.

HOW SHOULD THIS BOOK BE READ?

Audio Metering can be used as a reference book. The beginning contains a table of contents and the end of the book provides a glossary and an index.

Reading the book from cover to cover is highly recommended. The subject matter of the book has been organized so that the most basic material is placed at the beginning, while the more generally descriptive material is found towards the end of the book.

Enjoy

Eddy Bøgh Brixen
Smørum, Denmark, September 2010

Thanks to my family and to my publisher for the patience they have displayed during the preparation of this 2nd edition of Audio Metering. Also thanks to the publisher of the first edition, Broadcast Publishing, for leaving the rights to the author.

Chapter | one

Acoustic Sound

CHAPTER OUTLINE

"Sound" is an English-based word. "Audio" is derived from Latin, and refers to things that are related to hearing. Sound does not necessarily have to be audible. Infrasound and ultrasound, which are below and above the normal range of human hearing, respectively, are examples of inaudible sounds.

In English, we tend to use the terms "sound" and "audio" indiscriminately. When discussing acoustical topics in this book, we will be dealing with "sound," and that is where we will begin.

WHAT IS SOUND?

Sound is normally understood to mean elastic molecular oscillations in air or other media such as water, iron, or concrete. These oscillations result in pressure variations that are of such a magnitude that they can be sensed by human hearing.

However, sound can also be converted to, for example, variations in the electrical current in a conductor, or magnetic variations on an audio tape, or a sequence of numeric values. We call these forms **intermediate formats**, because we later convert them into acoustic sound.

SPEED OF SOUND

Sound propagates by an oscillating solid body setting the particles next to it in motion, and those next to them, and so on. Sound thus spreads with a certain

Audio Metering. DOI: 10.1016/B978-0-240-81467-4.10001-2

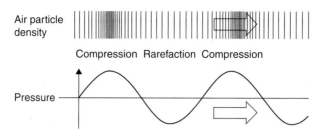

Air particle density

Compression   Rarefaction   Compression

Pressure

FIGURE 1.1 Sound (in air) can be defined as variations in air pressure.

propagation velocity. This is called the **speed of sound**, which varies depending on the medium.

The speed of sound (**c**) in a medium is normally specified in meters per second [**m/s**] or feet per second [**ft/s**]. In the air the speed of sound is dependent on the temperature:

In air, at 0°C or 32°F, the speed of sound is 331.4 m/s or 1086 ft/s (1193 km/t or 740.5 mph).

In air, at 20°C or 68°F, the speed of sound is 343.54 m/s or 1126 ft/s (1237 km/t or 767.7 mph).

However, as a rule 340 m/s or 1130 ft/s are used as approximations for the speed of sound for general purposes. Thus in applied audio technology, one often encounters the following values:

34 cm/ms (Sound travels about 34 cm each millisecond.)

3 ms/m (Sound takes roughly 3 ms to travel 1 meter.)

or

1 ft/ms (Sound travels about 1 ft each millisecond.)

FREQUENCY

Frequency (**f**) is a measure of the number of oscillations or cycles per second, and is specified in **hertz** (abbreviated as **Hz**).

$$f = \frac{1}{T} \; [Hz]$$

where
f = frequency [Hz]
T = period [s]

A frequency of 1 Hz = 1 cycle per second. A frequency of 1000 Hz = 1000 cycles per second (1000 Hz is normally expressed as 1 kiloHertz, abbreviated as 1 kHz).

The nominal audible frequency range of human hearing comprises frequencies from 20 Hz to 20 kHz. This range is called the **audio frequency range**.

WAVELENGTH

The wavelength is the distance a single oscillation takes to complete in a given medium. The wavelength is thus dependent on the frequency and the speed of sound in the medium concerned.

It can be expressed in the following manner:

$$\lambda = \frac{c}{f}$$

where
λ = wavelength [m]
c = speed of sound [m/s]
f = frequency [Hz]

or

λ = wavelength [ft]
c = speed of sound [ft/s]
f = frequency [Hz]

This relationship shows that the audio spectrum in air contains wavelengths ranging from approximately 17 m at 20 Hz down to 17 mm at 20 kHz or from approximately 55 ft at 20 Hz down to 0.055 ft at 20 kHz. This is referred to as the "wave theory boundary area," because the sound wavelengths are in the same order as the dimensions of the rooms and equipment that are used for recording and listening (microphones, loudspeakers, etc.).

This boundary area has an influence on everything from the acoustics of rooms to the directional patterns of loudspeakers, as well as the frequency response of microphones.

If the concept of wavelength is a little difficult to comprehend, it helps to imagine the sound radiating from a sound source as pressure variations generated by the source being carried away on a conveyer at a given fixed speed (the speed of sound). If one "freezes" this image the distance between two maxima is the wavelength of that frequency. If the frequency is increased, one period of that frequency is finished in less time and the travelled distance gets shorter, hence a shorter wavelength.

TABLE 1.1 Wavelengths in Air. In this Calculation the Speed of Sound is 344 m/s or 1126 ft/s.

Frequency		Wavelength		
20 Hz	⇒	17.20 m	or	56.3 ft
100 Hz	⇒	3.44 m	or	11.3 ft
200 Hz	⇒	1.72 m	or	5.65 ft
344 Hz	⇒	1.00 m	or	3.273 ft
1000 Hz	⇒	34.40 cm	or	1.126 ft
1126 Hz	⇒	30.20 cm	or	1.000 ft
2000 Hz	⇒	17.20 cm	or	0.563 ft
10,000 Hz	⇒	3.44 cm	or	0.113 ft
20,000 Hz	⇒	1.72 cm	or	0.056 ft

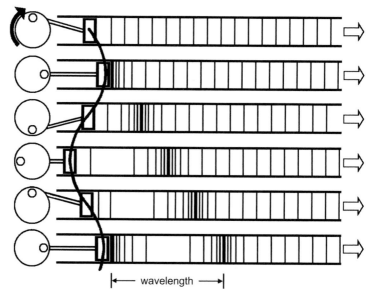

FIGURE 1.2 A sound source regarded as a piston. Pressure generated at the piston is transported away at a constant speed. Hence, the wavelength of a given frequency is the distance between two identical points in adjacent cycles.

SOUND PRESSURE

Air surrounds us. We call the static pressure of air the **barometric pressure**. Sound pressure is defined as the difference between the instantaneous pressure and the static pressure.

Pressure is measured in **pascals** (abbreviated **Pa**). One pascal is defined as 1 N/m^2 (newton per square meter).

The sound pressure can be regarded as a modulation of the static pressure. The greater the variations in sound pressure, the stronger the perception of the sound, i.e., loudness.

The weakest audible sound at 1 kHz has a sound pressure of approximately 20 µPa (20 micro-pascal $= 20 \cdot 10^{-6}$ Pa), whereas the ear's threshold of pain lies at a sound pressure of around 20 Pa. The weakest audible sound and the threshold of pain thus differ by a factor of one million.

CONVERSION RELATIONSHIPS

The pascal (Pa) is a unit of the SI system (International System of Units). Other units are still seen in practice, for example on older data sheets for microphones, where the bar unit is used.

SOUND POWER

Sound is a form of energy, hence the concept of sound power.

Sound power is the sound energy transferred during a period divided by the period of time concerned. In a travelling plane wave with a sound pressure

TABLE 1.2 Conversion Between Pa, Bar, and Atmosphere (Atm).

Unit	Equivalent to:
1 Pa	1 N/m^2 $10 \text{ } \mu\text{bar}$ $7.5006 \cdot 10^{-3} \text{ mm Hg}$ $9.869 \cdot 10^{-6} \text{ atm}$
bar	10^5 N/m^2 10^5 Pa 750.06 mm Hg 0.98692 atm
1 atm	$1.01325 \cdot 10^5 \text{ N/m}^2$ $1.01325 \cdot 10^5 \text{ Pa}$ 1.01325 bar 760 mm Hg

of 20 μPa, the power that passes through an area of 1 m^2 placed perpendicular to the direction of travel is

1 pW (one pico-watt $= 10^{-12}$ watt).

This value is used as a reference when specifying a sound power level. For a sound pressure of 20 Pa, the power is 1 W.

Sound power, for example from a loudspeaker or from a machine, is measured in practice either in a reverberation chamber (a room with highly reflective surfaces) or by performing a large number of sound pressure measurements around the object; these measurements subsequently are used to calculate the power.

SOUND INTENSITY

Sound intensity is an expression of power per unit of area. For a flow of sound energy that is propagating in a specific direction, the intensity is the power that is transferred through an area perpendicular to the direction of travel, divided by the area.

For a travelling plane wave with a sound pressure of 20 μPa, this intensity is 1 pW/m^2.

Sound intensity has a directional component. In practice, the sound intensity from a sound source is measured using a sound intensity probe consisting of two transducers (pressure microphones) at a well-defined distance. By looking at the phase of the radiated sound, among other things, the direction of the sound can be determined.

SOUND FIELDS

Spherical Sound Field

When sound is radiated from a point source, the intensity decreases with distance. It can be compared with a balloon when being filled with air: the

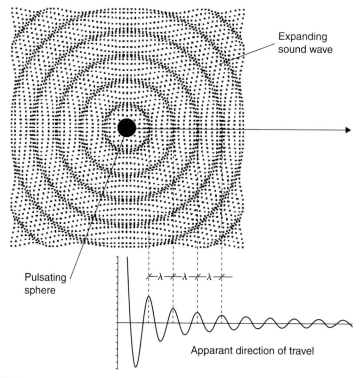

FIGURE 1.3 Sound radiating from a point source.

bigger the diameter, the thinner the rubber wall at a given area on the circumference. The balloon consists of a certain amount of rubber that has to cover a still larger volume. There is a quadratic relation between the radius and the area. Doubling the radius enlarges the area by four.

The intensity falls with distance according to:

$$I = \frac{1}{d^2}$$

where
I = sound intensity
d = distance

This is called the **inverse square law**.
Therefore, the sound pressure falls according to:

$$P = \frac{1}{d}$$

where
P = sound pressure
d = distance

Thus, doubling the distance halves the sound pressure. In dB, the sound level drops 6 dB when doubling the distance.

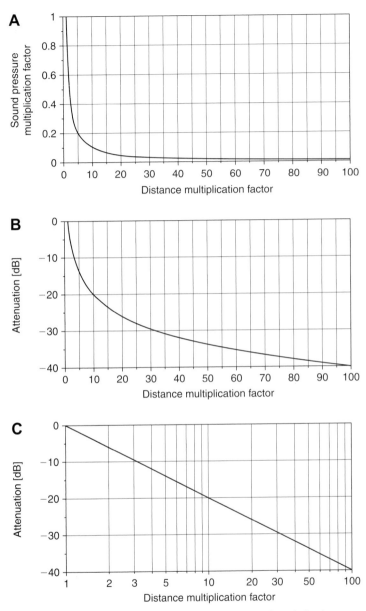

FIGURE 1.4 Point source: Three different ways to express the relation between sound pressure and distance. A: Sound pressure (expressed as a multiplication factor) vs. distance (expressed as a multiplication factor). Example: Moving 10 times the initial distance (meters, feet, yards,…) away from the sound source changes the sound pressure by a factor of 0.1. B: Attenuation of sound pressure (expressed in dB) vs. distance (expressed as a multiplication factor). Example: Moving 10 times the initial distance away from the sound source reduces the sound pressure by 20 dB. C: Sound pressure (expressed in dB) vs. distance (expressed as a multiplication factor on a logarithmic scale). The values are the same as in B. However, the curve here is a straight line.

Cylindrical Sound Field

An (infinite) line source generates a cylindrical sound field.
The intensity falls with distance according to:

$$I = \frac{1}{d}$$

where
I = sound pressure
d = distance

Hence the sound pressure falls according to:

$$P = \frac{1}{\sqrt{d}}$$

where
I = sound pressure
d = distance

Thus, doubling the distance reduces the sound pressure by $\sqrt{2}$. In dB, the sound level drops 3 dB when doubling the distance.

Some loudspeaker designs are pronounced line sources, typically formed by an array of speaker units. In practical audio engineering these loudspeakers may act as line sources within a limited distance and frequency range. In the far field the practical (finite) line array is always considered to be a point source.

Plane Sound Field

In a plane sound field the sound does not attenuate as the intensity of the sound field is kept constant in the direction of the propagation. In other words, the attenuation is 0 dB. However, infinite plane sound sources do not exist. In real life an approximation of the plane sound field can be experienced far away from a point source where a limited sector of a spherical sound field can be regarded as plane. Another approximation exists very close to a large loudspeaker membrane or a vibrating wall. (For instance, if a microphone is placed 1 cm from a 12-inch loudspeaker unit and then moved 1 cm further away it does not really change the sound level.)

Chapter two

From Acoustic Sound to Electrical Signals

CHAPTER OUTLINE

In order to be able to measure, manipulate, or describe sound, we generally have to convert it from an acoustic to an electrical signal.

Sound exists in purely acoustical terms as pressure variations in the air. By using an appropriate transducer, for example a microphone, these pressure variations can be transformed into variations in current or voltage.

If we wish to record the sound, we then have to convert these electrical variations into another form. In the past these variations were widely used to leave magnetic traces on an audio tape or physical variations in the surface of a vinyl record. Today we mostly convert the sound into sequences of numerical values, as is the case with digital technology. Computer technology can subsequently be used to save or process these values, which presumably will end up as acoustic information again at a later date.

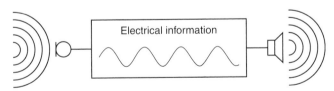

FIGURE 2.1 From acoustic sound to electrical signal and back to acoustic sound.

ELECTRICAL SIGNALS

When a microphone is used to convert acoustic information to electrical information, we say that the electrical signal is analogous to the acoustic signal. In other words, the waveforms that describe the pressure and voltage or current variations resemble each other.

In the electrical world we have a circuit in which a current of electrons flows. A potential difference drives the electrons in one direction or the

Audio Metering. DOI: 10.1016/B978-0-240-81467-4.10002-4

other. It is the size of this potential difference, the voltage, which we most often regard as the magnitude of the signal.

SPEED

The movement of the electrons occurs at a speed close to the speed of light, which in a vacuum is $2.99792458 \cdot 10^8$ m/s [$9.8293 \cdot 10^8$ ft/s]. In practice, we use a rounded value of $3 \cdot 10^8$ m/s, or 300,000 km/s [186,000 miles/s].

Clearly, electrical signals moves at a much higher speed in a wire than acoustic signals propagate in air.

WAVELENGTH

The wavelength of the electrical signal in cables and transmission lines is determined by the same expression that is used for acoustic sound:

$$\lambda = \frac{c}{f}$$

where

λ = wavelength [m]
c = speed [m/s]
f = frequency [Hz]

or

λ = wavelength [feet]
c = speed [feet/s]
f = frequency [Hz]

In the audio frequency range (20 Hz − 20 kHz) we thus have wavelengths ranging from 15,000 km down to 15 km or from 9000 miles down to 9 miles. As opposed to acoustic sound, which propagates in the air, very seldom will audio converted to electrical signals have wavelengths that are of the same size as the circuits we are working with. In practice, this will for the most part only occur when we are using cables with lengths in kilometers or miles.

If, however, we are working in the high frequency spectrum, for example with video or radio frequencies, or with transmission of digital signals, then the wavelengths will very quickly turn out to be comparable with the sizes of the physical circuits in which the current is flowing.

Digital Representation

CHAPTER OUTLINE

Digital audio technology is based on the transformation of a signal varying in an analog manner to numerical values at an appropriate rate. After this conversion, computers can be used in the processing, transmission, and storage of the signals.

Other advantages of digital audio technology include error correction, which allows for the execution of copying and transmission in a lossless manner. In addition it enables what otherwise would have to be done with physical components such as resistors, capacitors, and inductors to now be represented as simple calculations.

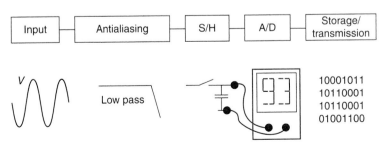

FIGURE 3.1 Principles for the digitizing of analog signals. The signal is low-pass filtered before a sampling of the signal is performed. The magnitude (quantization) of each sample is then determined. The resolution is determined by the number of bits.

Audio Metering. DOI: 10.1016/B978-0-240-81467-4.10003-6

ANTIALIASING

Before the analog signal can be converted to a digital signal, it is necessary to determine a well-defined upper cut-off frequency (f_u), and a low-pass filter is used for this. This filtering is called **antialiasing**; the term "alias" means an assumed identity. The necessity of the filtering is due to the sampling process itself. The analog signal must not contain frequencies that are higher than half of the sampling frequency (a frequency also called the **Nyquist frequency**). If the sampling frequency is lower than twice the highest input frequency, then the reconstructed signal will contain frequency components that were not present in the original. The filter ensures that the signal does not contain any aliasing frequencies after reconstruction.

SAMPLING

After the low-pass filtering, sampling is performed. Sampling consists of measuring the instantaneous value of the signal. The frequency at which this measurement is taken is called the **sampling frequency** (f_s).

A comparison can be made with a movie camera that can record moving pictures by taking a single picture 24 times per second. One could then say that the camera's sampling frequency is 24 Hz. Now and then you can also observe alias frequencies elsewhere, such as when we see the wheels of the stagecoach turning backwards while the horses and the carriage are moving forward.

Sampling frequencies of 32 kHz, 44.1 kHz, and 48 kHz have long been the standard for quality audio for things like CD or broadcast audio tracks.

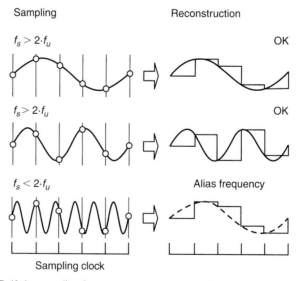

FIGURE 3.2 If the sampling frequency is not at least twice the highest audio frequency the reconstructed signal will not be in accordance with the input.

However, the use of 88.2 kHz, 96 kHz, 176.4 kHz, and 192 kHz has gradually also become commonplace. The latter are seen in use particularly with DVD and Blu-ray audio tracks.

Sound clips for computer games, audio in communication systems, and other similar types of audio typically use very low sampling frequencies down to 8 kHz or even less.

For each sample, the instantaneous value of the analog signal is retained for as long as the analog to digital converter (also called an **A-D converter** or **ADC**) needs to perform its conversion. In the early converters this was performed by a "hold" circuit, which fundamentally was a capacitor that was charged/discharged to the instantaneous value of the signal at the point in time the sample was taken. The reading of the analog signal in modern converters occurs so quickly that the hold function can be omitted. However, the understanding of the sampling process is easier when keeping a "virtual capacitor" in mind.

Oversampling, sampling done at a frequency that is a number of times higher than the requisite minimum, is performed in many converters. Oversampling is utilized because it makes it easier to implement antialiasing filters. In addition, oversampling is a necessity when the signal must be resolved into many bits, again because it is not possible to implement filters that are as sharp as would be needed to, for example, be able to make a difference at a resolution of 24 bits.

The SACD (Super Audio Compact Disc) uses oversampling providing a direct stream of data that requires a sampling frequency 64 times that of the standard CD and ends up with a sampling frequency of 2.8224 MHz.

QUANTIZATION

Now comes the part of the process that determines the digital "number." This process is called **quantization**. The word comes from Latin (*quantitas* = size). During quantization, the size of the individual sample is converted to a number. This transformation, or conversion, is not always completely ideal, however.

The scale that is being used for purposes of comparison has a finite resolution that is determined by the number of bits. The word "bit" is a contraction of the words "binary digit," which refers to a digit in the binary number system. With quantization, it is the number of bits that determine the precision of the value read. Each time there is one more bit available, the resolution of the scale is doubled and so the error in measurement is halved. In practice, this means that the signal-to-noise ratio is improved by approx. 6 dB for each extra bit that is available.

BINARY VALUES

The value ascribed to the quantization is not a decimal number but, rather, a binary number. The binary number system uses the number 2 as its base

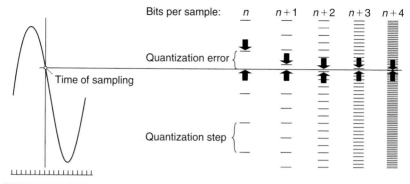

FIGURE 3.3 With quantization, it is the number of bits that determine the precision of the value read. Each time there is one more bit available, the resolution of the scale is doubled and the error in measurement is halved. In practice this means that the signal-to-noise ratio is improved by approximately 6 dB for each extra bit that is available.

number. This means that only two numbers are available, namely 0 and 1. These values are easy to create and detect in electrical terms. For example, there is a voltage (1), or there is not a voltage (0); the current is running in one direction (1), or the current is running in the opposite direction (0).

With one digit, or one bit, available we thus only have two values, namely 0 and 1. With two bits available, we have four possible combinations, namely 00 (zero, zero), 01 (zero, one), 10 (one, zero) and 11 (one, one). The number of steps on the scale equals the number of bits to the power of two. In practice, between 8 and 24 bits are used in the quantization of analog signals. CD-quality audio corresponds to 16 bits per sample ($= 16^2 = 65,536$ possible values). There are only a finite number of values available when the magnitude of the signal is determined. This means that the actual analog value at the moment of sampling is in fact represented by the nearest value on the scale.

With linear quantization (equal distance between the quantization steps), a resolution of only a few bits would result in extreme distortion of the original signal. When it is resolved with additional bits, this distortion gradually becomes something that can be perceived as broadband noise. As a rule of thumb, the signal-to-noise ratio is estimated to be about 6 dB per bit.

A-D

The principal components in the A-D converter are one or more comparators, which compare the instantaneous values of the individual samples with a built-in voltage reference. After the comparison, the comparator's output will indicate the value 0 (or "low") if the signal's instantaneous value is less than the reference. If the signal's instantaneous value is equal to or greater than the reference, then the output of the comparator will indicate the value 1 (or "high").

For serial (sequential) quantization, the comparator will first determine the most significant bit, and then the next bit, etc. until the least significant bit has been determined. For a parallel conversion, a comparator is required for each level that is to be determined, which for n bits corresponds to 2^{n-1}. If, for example, there are eight bits available for the total signal, this corresponds to a resolution of 256 levels, represented by numerical values in the range 0-255. Written in binary format, this corresponds to the numbers from 00000000 to 11111111. A form of encoding is normally used where the first digit specifies the polarity of the signal. If the number is 0 then a positive voltage value has been sampled. If the number is 1, then a negative voltage value has been sampled.

Many converters have been designed according to the Delta-Sigma principle. This uses oversampling with a frequency so high that it only needs to be determined for each sample whether the current value is greater or smaller than the prior value. The advantage is that errors can only arise of a magnitude corresponding to that of the smallest quantization interval, whereas the errors in parallel conversion can be much greater. After the conversion, the long sequence of serial information can be reorganized into a standard parallel bit format at a standardized sampling frequency, so that it can be used for CD, DAT, etc. For SACD, the bit stream (Direct Stream Digital or DSD) generated by the Delta-Sigma converter is what is recorded.

Some converter types combine parallel and serial conversion (Flash converter), where four or five bits are typically determined at a time. This combines high speed with good precision.

D-A

In the conversion from digital to analog, the objective is to produce a signal that is proportional to the value that is contained in the numerical digital information. This can be done in principle by having each bit represent a voltage source such that the most significant bit is converted into the largest voltage, the next most significant bit is converted into half of that voltage, etc. All of

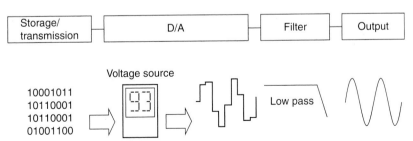

FIGURE 3.4 During digital-to-analog conversion, the stored numbers are converted back to an analog signal. The numbers are essentially read into a programmable power supply, so that they re-create the corresponding voltage steps. The low-pass filter smoothes out the signal by removing the harmonic overtones (caused by the steps) lying above the desired frequency spectrum.

the voltage steps are added, and a holding circuit ensures that the signal is continuous until the next sample has been reestablished. The signal created is then smoothed out by the use of a low-pass filter.

The D-A conversion is in principle quite simple; however it can be difficult to control in the real world, where for example $2^{16} = 65,536$ different levels could be generated for a 16-bit signal. There can certainly be differences in the quality of A-D converters in practice. Poor converters can have a DC offset and poor linearity in their dynamics. Methods exist however, to reduce these problems.

BIT REDUCTION

The quality of digital sound can in principle be determined by the number of bits per sample and by the sampling frequency. In both cases, the higher the better. The problem is that for many purposes, transmitting sound over the Internet, storage for handheld devices, etc., it is not possible to transfer the number of bits per seconds required for high-quality audio (i.e., CD, SACD, DVD, Blu-ray) within a reasonable amount of time.

Therefore some compromise must be introduced, such as the number of bits per second being lowered. This is called **bit reduction** or **bit companding** (a mixture of the words compressing and expanding). Fundamentally, there are a number of different methodologies available.

Lossless Packing

One principle for reducing the number of bits does not actually throw any information away. One system is known as **MLP, Meridian Lossless Packing**. This is equivalent to zipping a data file. The information is packed so it takes up less space but the contents are still intact. Another system is **FLAC, Free Lossless Audio Codec**. which is very popular due to its fast decoding. As it is a non-proprietary format several codecs are available. The store data are reduced to approximately half size.

Lower f_s and Fewer Bits per Sample

The simplest method is to use a lower sampling frequency and fewer bits per sample; however, this results in deterioration in quality.

Nonlinear Quantization

A method that has been used for many years is **nonlinear quantization**. with specifically the A-law (telephony in Europe) and μ-law (mu-law, telephony in the US) methods being the most widely used variants. These require only 8 bits per sample but effectively give 12 bits of resolution basically obtained by fine resolution at low levels and an increasingly more coarse resolution as levels gets higher. This method is often used in communications; however, the quality is not good enough for music.

Perceptual Coding

The dominating methodology is called **perceptual coding**, and is based on psycho acoustics. It makes use of the fact that the ear does not necessarily hear everything in a complex spectrum. Strong parts of the spectrum mask weaker parts. The principle is then that what is not audible can be discarded. (Read more about masking in Chapter 7.)

For perceptual coding, a frequency analysis is performed. One single sample by itself has no frequency information; hence a greater number of samples are collected, typically 1024. Calculations are then performed from frequency band to frequency band determining whether signals in the surrounding parts of the frequency spectrum are masking precisely this band. The data in bands that are masked are more or less thrown away. In addition, multiple channels can share information they have in common. Bits are only used in those ranges that are most important for the sequence concerned. Depending on the algorithms used the contents may be reduced to a few percent of the original size.

One of the drawbacks of all these methodologies is that it takes time to compress the bit stream and it takes time to expand it again. Time delays of up to a few hundred milliseconds will be experienced in the transmissions, solely due to the complexity of the algorithms. With perceptually coded signals, another problem can arise when any kind of signal processing is applied. The thresholds that might have kept the artifacts at an audible minimum suddenly may change and have an influence on the sound quality perceived.

Codecs and Applications

There are an overwhelming number of bit reduction algorithms available. Some are initiated by standards organizations while others are proprietary company standards. The different methods are in general optimized for different applications like download and storage for personal playback devices, Internet media, VoIP, video embedded audio, digital broadcast, etc. Often new algorithms are

TABLE 3.1 The Most Popular Non-Proprietary Formats for Perceptually Coded Audio.

Codec	bit rates kbps	sample rates kHz	filename extension
MP3 (MPEG 1, Layer 3)	32–320	32, 44.1, 48	mp3
MP3 (MPEG 1, Layer 3)	8–160	16, 22.05, 24	mp3
AAC (Advanced Audio Coding)	variable and dependent on no. of channels (up to 48 channels) Hi quality stereo: 128 kbps Hi quality 5.1: 320 kbps	8, up to 96	m4a, m4b, m4p, m4v, m4r, 3GP3gp mp4, aac

based on older versions and may or may not be backward compatible. This is an area of constant development. So the following compressed overview in Table 3.1 can be regarded as a snapshot providing information on a few currently widely used algorithms.

HOW MUCH SPACE DOES (LINEAR) DIGITAL AUDIO TAKE UP?

When calculating the size of any digital information handled by computers one has to be aware that it is all based on bytes [B], which each contain 8 bits. This is why the number of bits per sample is calculated as an integer times the number 8 ($1 \cdot 8$, $2 \cdot 8$, $3 \cdot 8$, etc). The number of bits per sample of linear PCM (Pulse Code Modulation, digital audio) is basically 8 (1 byte), 16 (2 bytes), 24 (3 bytes), or 32 (4 bytes). For internal processing 64 bits or more can be used.

Because these numbers get large the use of prefixes gets very handy. Here we use "k" (kilo), "M" (Mega), "G" (Giga), "T" (Tera), etc. The sizes are calculated as follows:

$$1 \text{ kB} = 1024 \text{ B} = 8192 \text{ bits}$$
$$1 \text{ MB} = 1024 \text{ kB} = 8{,}388{,}608 \text{ bits } (\approx 8.39 \cdot 10^6 \text{ bits})$$
$$1 \text{ GB} = 1024 \text{ MB} \approx 8.59 \cdot 10^9 \text{ bits}$$
$$1 \text{ TB} = 1024 \text{ GB} \approx 8.8 \cdot 10^{12} \text{ bits}$$

Example:

How much storage capacity is needed for a 1-hour stereo recording in 44.1 kHz/16 bit?

The total number of bits is calculated as follows:
Sampling frequency·no. of bits per sample·no. of audio channels·the duration of the recording:
44,100 (samples per second)·16 (bits per sample)·2 (channels)·1 (hour)· 60 (minutes)·60 (seconds) = $5.08 \cdot 10^9$ bits

Number of bytes: $5.08 \cdot 10^9 / 8 = 6.35 \cdot 10^8$ B
Number of kB: $6.35 \cdot 10^8 / 1024 = 6.20 \cdot 10^5$ kB
Number of MB: $6.20 \cdot 10^5 / 1024 = 605.6$ MB

This is in the range of the storage capacity for a CD. For this example, additional data such as the file header and table of contents, is not taken into consideration.

Chapter | four

Signal Types

CHAPTER OUTLINE

Sound is pressure that varies over time. Represented in electrical form, it is voltage or current that varies over time. This means that every sound — regardless of how many frequencies it might contain — can be described by a signal's variation in time or waveform. Some of the more typical waveforms and their frequency content are discussed in this chapter.

PURE TONES

Pure tones are characterized by a sinusoidal waveform (see Figure 4.1). These are periodic signals that contain one and only one frequency. In practice, pure tones almost always occur as test signals only.

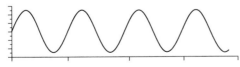

FIGURE 4.1 A pure (sinusoidal) tone contains one and only one frequency.

COMPLEX TONES

Complex tones are periodic signals that are based on a fundamental frequency with associated harmonic overtones. These signals are, among other things,

Audio Metering. DOI: 10.1016/B978-0-240-81467-4.10004-8

characteristic of musical instruments. In musical contexts, overtones are referred to as **partials** or **harmonics**. When one refers to the frequency of a tone, what they are actually referring to in reality is the frequency of the fundamental frequency.

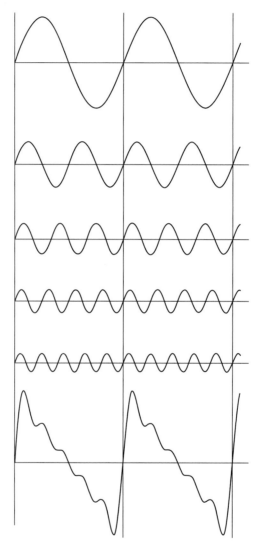

FIGURE 4.2 Five pure tones are shown here, corresponding to a fundamental tone, the associated second harmonic overtone (with an amplitude $= 1/2 \cdot$ the fundamental tone), a third harmonic overtone (with an amplitude $= 1/3 \cdot$ the fundamental tone), a fourth harmonic overtone (amplitude $= 1/4 \cdot$ the fundamental tone), and a fifth harmonic overtone (amplitude $= 1/5 \cdot$ the fundamental tone). At the bottom, these five pure tones are combined. The waveform approximates a saw tooth waveform. Note that all the curves here start at $0°$. If the phase relationship (the phase angle) were different, then the resultant waveform would also have a different appearance.

The frequency of the harmonic overtones will be integer multiples of the frequency of the fundamental tone. If the fundamental tone is 100 Hz, then the second harmonic (second overtone) would be 200 Hz, the third harmonic 300 Hz, the fourth harmonic 400 Hz, etc.

SPECIAL WAVEFORMS

Among the periodic waveforms, there are typical waveforms such as a sinusoid, saw tooth, triangular, square, and pulse train (or pulse string). All of them have more or less characteristic frequency content.

The **saw tooth wave** is characterized by the fact that it contains the fundamental frequency and all harmonics in a specific proportion (the second harmonic has half the amplitude of the fundamental, the third harmonic has a third the amplitude of the fundamental, etc.).

The **square wave** is characterized by the fact that it contains the fundamental frequency and all the odd numbered harmonics in the same declining proportion as the saw tooth wave.

The square wave is known from, among other places, electro-acoustic equipment that is overloaded and thus "clips" the peaks off the signal. This is called **harmonic distortion**. The magnitudes of the harmonic overtones created are stated in relation to the fundamental tone and are expressed as percentages (i.e., total harmonic distortion (THD) equals $xx\%$).

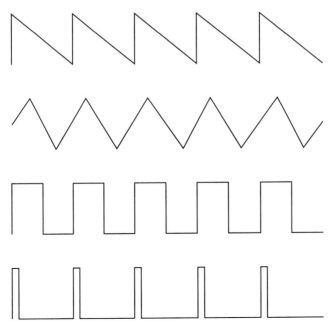

FIGURE 4.3 Special waveforms: Sawtooth, triangular, square and pulse train.

NOISE SIGNALS

Those signals for which no sensation of tone occurs are called **noise**. These are characterized by, among other things, all frequencies in a given range of frequencies being represented. Some well-defined electrical noise signals belong to this group, and are used as test signals:

White noise is a signal that contains constant energy per Hz bandwidth.

Pink noise is a signal that contains constant energy per octave (or 1/3 octave). The amplitude follows the function 1/f, which corresponds to it diminishing by 3 dB per octave or 10 dB per decade.

Brown noise is a signal where the amplitude diminishes by $1/f^2$, corresponding to 6 dB per octave or 20 dB per decade.

Real noise signals such as traffic noise or ventilation noise can, in addition to broadband noise, also contain audible tones. The spectral distribution may show that the primary content is found in a specific part of the frequency spectrum. For example, ventilation noise contains a primary content of low frequencies, and compressed air noise has a primary content in the high frequencies. Noise signals can also be a part of the sound of musical instruments, for example the "resin sound" of the strings or the air noise of various wind instruments.

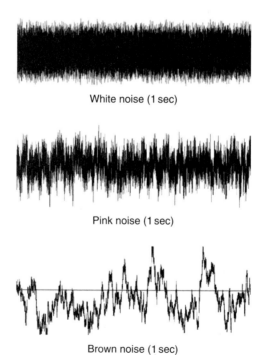

White noise (1 sec)

Pink noise (1 sec)

Brown noise (1 sec)

FIGURE 4.4 Waveforms of white, pink, and brown noise (1-second excerpts).

THE VOICE AS A SOUND SOURCE

It is important to understand the characteristics of the sound of voice. While language can be something that groups of people have in common, the sound and character of the voice is unique from person to person. Our familiarity with speech, regarded as an acoustic signal, allows us to have a good reference as to how it should sound.

Sound Level

The level of the voice can vary from subdued to shouting. This level is naturally individual from person to person and thus difficult to assign a number to. The values in Table 4.1 represent the average sound level of speech of an adult.

The ability to understand speech is optimum when the level of the speech corresponds to normal speech at a distance of 1 meter, in other words a sound pressure level of approximately 58 dB re 20 μPa. ("re" means "with reference to." Here, it is with reference to the weakest audible sound pressure level. See Chapter 6 on dB.)

The Spectrum of Speech

The spectrum of speech covers a large part of the total audio frequency range. Speech (Western languages) consists of vowels and consonant sounds. The voiced sounds in speech are generated by the vocal chords and then influenced by the resonances of cavities they pass through. Speech can be regarded as a fundamental frequency with a large number of harmonics. A whisper does not contain voiced sounds; however, the cavities that contribute to the formation of the different vowels will still act on the passing flow of air. This is how the characteristics of vowels may occur in a whisper.

TABLE 4.1 Average Speech Level as a Function of Listening/Recording Distance. There is an Approximately 20 dB Difference Between Normal Speech and Shouting.

| Listening distance [m] | Speech level [dB re 20 μPa] | | | |
	Normal	Raised	Loud	Shout
0.25	70	76	82	88
0.5	65	71	77	83
1.0	58	64	70	76
1.5	55	61	67	73
2.0	52	58	64	70
3.0	50	56	62	68
5.0	45	51	57	63

In general, the mean value of the fundamental frequency — also called the **pitch** or f_0 — is in the range of 110–130 Hz for men, and approximately one octave higher at 200–230 Hz for women. In both cases values outside these ranges occur. For children, f_0 lies at around 300 Hz. The individual deviation from the base frequency is higher in tone languages (like Chinese) than in European languages like English or German. Also, the mood of the speaker affects the pitch.

The consonants are formed by air blockages and noise sounds created by the passage of air in the throat and mouth, and particularly the tongue and lips. In terms of frequency, the consonants lie mostly above 500 Hz.

At a normal vocal intensity, the energy of the vowels normally diminishes rapidly above approximately 1 kHz. Note, however, that the emphasis in the speech spectrum shifts one to two octaves towards higher frequencies when the voice is raised. One should also note that it is not possible to increase the sound level for consonants to the same extent as for vowels. In practice this means that the intelligibility of speech is not increased by shouting in comparison to using a normal voice assuming the background noise is not significant. It is also worth knowing when recording speech that the speech spectrum changes with distance and angle.

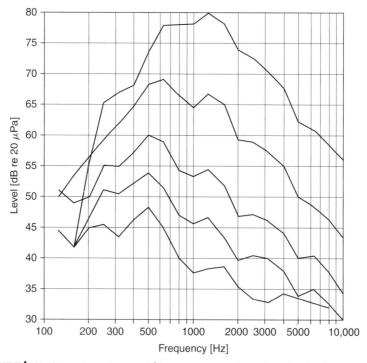

FIGURE 4.5 Male voices: Average 1/3-octave spectra for various levels of speech. Note that the energy is moved towards higher frequencies as the sound level of the voice is gradually increased.

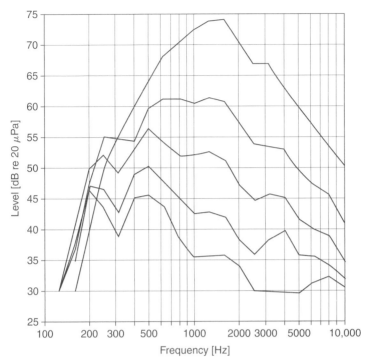

FIGURE 4.6 Female voices: Average 1/3-octave spectra for various levels of speech.

Formants

If one listens to two or more people who are speaking or singing the same vowel at the same pitch (f_0), the correct vowel is presumably recognized in all cases, although there will be a difference in terms of timbre.

The characteristics of the individual vowels are formed by the acoustic filtering of the signal from the vocal chords. The result is a number of **formants**, i.e., frequency ranges that are particularly prominent, which give the sound of the vowels, more or less regardless of the pitch of the voice. This sound-related formation of vowels is something that is fundamentally learned by the talker. In contrast, there will be frequency-related characteristics in the formant ranges of

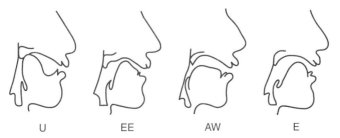

FIGURE 4.7 The position of the tongue is shown here for four vowels.

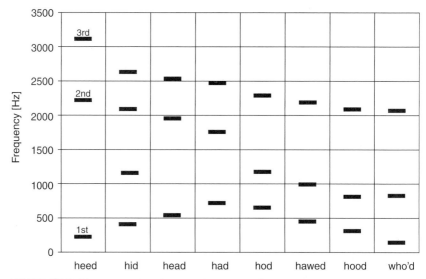

FIGURE 4.8 Placement of the formants for different vowels in English. The formants are marked from the bottom (lowest frequency) as 1st formant, 2nd formant, and 3rd formant.

individual voices that are due to the individual anatomical differences from person to person. These differences are, along with other phenomena such as intonation, part of what makes it possible to differentiate one voice from another.

Crest Factor

The **crest factor** expresses the relation between the peak level and the RMS level (see Chapter 5). The consonants can have relatively strong peaks, but limited energy content when viewed over a longer period of time. One can convince oneself of this by comparing the amplitude reading on a fast meter like TP (true peak) or PPM (peak programme meter) and a slower meter like VU (volume indicator) or LU (loudness unit), respectively. The crest factor is typically 10–15 (20–23 dB). This has significance when a voice is recorded or reproduced in an electro-acoustic system.

MUSICAL INSTRUMENTS

The sound of a musical instrument can be characterized by parameters such as pitch, tone, timbre, range of harmonics, overtones, attack, decay, and formants.

Tone

The tone of a musical instrument essentially consists of a fundamental frequency and a number of harmonics and overtones/partials. Characteristic noise sounds are added to this base.

The magnitude of the fundamental frequency may be very small compared to the rest of the spectrum. However, due to ears' ability to define the fundamental frequency based on the interval between harmonics the pitch is in general correctly perceived. It is important to distinguish between the definition of harmonics and overtones or partials when looking at the the acoustics of musical instruments. Harmonics are always a mathematical multiple of the fundamental frequency. However, overtones are not always necessarily harmonic; in this case they are called inharmonic overtones.

Depending on the type of instrument the tone will be characterized by attack and by decay. Each partial may actually have different attack and decay patterns. Percussion instruments do not normally exhibit a sustained tone; however, by the stroke a resonance is excited.

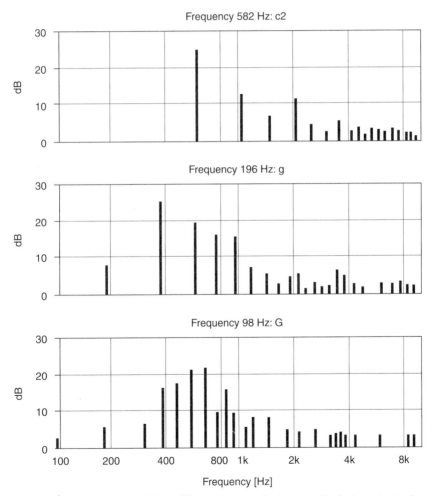

FIGURE 4.9 The spectra of three different notes on the bassoon. At the lowest note the fundamental frequency is almost not present. Notice the formant range, 300–600 Hz.

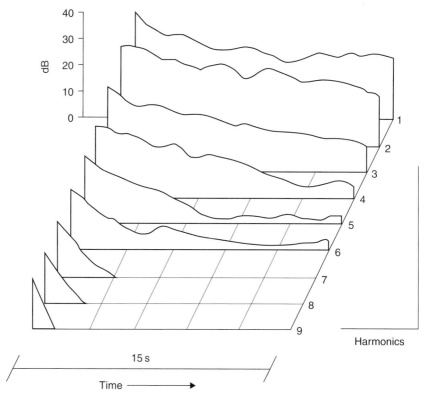

FIGURE 4.10 The build up, attack, and decay of individual harmonics (piano).

The noise that is attached as a part of the characteristic sound is common for almost all musical instruments. Examples of this are the sound of the air leakage on the mouthpiece, the sound of resin on the bowed string, and the sound of different instruments' moving mechanical parts.

Like the voice, many musical instruments also have formants, which are related to the physics of the instruments. For instance a bassoon consists of a tube of a given length. This "pipe" provides a natural emphasis of a given frequency range, in this case around 300–600 Hz (see Figure 4.9). No matter which note is played, the harmonics within this frequency band are always the strongest. The specific formants are regarded as the typical range of the instrument when filtered in a complex down-mix in order to avoid "mud" in the mix but still keep the characteristics of the individual instruments. In addition to the characteristics of the instrument itself, the musicians' personal style and playing techniques also obviously play a role.

ACOUSTIC MEASURES OF MUSICAL INSTRUMENTS

Table 4.2 contains some acoustical characteristics for a selection of musical instruments.

TABLE 4.2 Acoustical Properties of Musical Instruments

Instrument	Tonal range		Frequency content	Typical max SPL, 3 m	SPL	Dynamic range	Attack time
	Lower	Upper	Harmonics/Noise				
Strings							
Violin	g (196 Hz)	g^4 (3136 Hz)	⇒ 10 kHz / 16 kHz	95 dB	At the ear: up to 109 dB	30 dB	5–30 ms
Viola	c (131 Hz)	c^3 (1047 Hz)	⇒ 10 kHz / 16 kHz			30 dB	30–50 ms
Cello	C (65.4 Hz)	c^3 (1047 Hz)	⇒ 8 kHz / 16 kHz			35 dB	Plucked string: 5–30 ms; Struck string: 30–50 ms
Double bass	E_1 (41.2 Hz)	g (196 Hz)	⇒ 7.5 kHz / 10 kHz	95 dB		35 dB	Up to 110 ms; Hard: up to 110 ms; Soft: up to 450 ms
Brass							
Trumpet	e (165 Hz)	d^3 (1175 Hz)	⇒ 15 kHz / 15 kHz	106 dB	0.5 m from bell piece: normal forte: 108 dB; extreme forte: 128 dB	Lower tonal range: 30 dB; Higher tonal range: 10 dB	20–40 ms
French horn	H_1 (61.7 Hz)	f^2 (698 Hz)	⇒ 10 kHz / 10 kHz	95 dB		Middle tonal range: 40 dB; Higher tonal range: 20 dB	20–40 ms; In low tonal range: up to 80 ms
Trombone	E (82.4 Hz)	c^2 (523 Hz)	⇒ 7 kHz / 10 kHz	104 dB		Middle tonal range: 45 dB	20–40 ms

(Continued)

TABLE 4.2 Acoustical Properties of Musical Instruments—Continued

Instrument	Tonal range	Frequency content	Typical max SPL, 3 m	SPL	Dynamic range	Attack time
Tuba	$Eb1$ (39 Hz) g^1 (392 Hz)	⇒ 4 kHz / 7.5 kHz	96 dB		Middle tonal range: 40 dB	20—40 ms
Woodwind						Generally 20—60 ms
Piccolo flute	d^2 (587 Hz) c^5 (4186 Hz)	⇒ 10 kHz / 15 kHz	102 dB			
Flute	c^1 (262 Hz) c^4 (2093 Hz)	⇒ 6 k Hz / 15 kHz	96 dB		35 dB	Flute, low register: Up to 180 ms
Oboe	c^1 (262 Hz) d^3 (1175 Hz)	⇒ 15 kHz / 15 kHz	90 dB		30 dB	Oboe, low register: Up to 120 ms
Clarinet	d (147 Hz) e^3 (1319 Hz)	⇒ 10 kHz / 15 kHz	92 dB		50 dB	
Bassoon	Bb (58.3 Hz) c^2 (523 Hz)	⇒ 8 kHz / 12 kHz	90 dB		35 dB	
Contra bassoon	Bb_2 (29.2 Hz) c^1 (262 Hz)	⇒ 8 kHz / 12 kHz	92 dB			
Soprano sax	ab (208 Hz) eb^3 (1245 Hz)	⇒ 12 kHz / 16 kHz	98 dB			
Alto sax	db (139 Hz) ab^2 (831 Hz)		98 dB	At the center of the bell piece up to 130 dB		
Tenor sax	Ab (104Hz) eb^2 (622 Hz)		98 dB	At the center of the bell piece up to 130 dB	25 dB	
Baritone sax	Db (69.4 Hz) ab^1 (415 Hz)		98 dB			
Bass sax	Ab_1 (51.9 Hz) db^1 (278 Hz)		96 dB			

	Tuning	Frequency range with highest level			
Acoustic guitar (nylon strings)	E (82.4 Hz) c^3 (1047 Hz)		88 dB		
Acoustic guitar (steel strings)	E (82.4 Hz) c^3 (1047 Hz)		92 dB		
Grand piano 3 strings: c^5 – F 2 strings: F – F_1	A_2 (27.5 Hz) c^5 (4186 Hz)		Lid in 45° position: 100 dB		
Percussion	Tuning	Frequency range with highest level			
Timpani 80 cm 75 cm 65 cm 60 cm	Eb (77.8 Hz) – G (98 Hz) F(87.3) – c (131 Hz) A(110 Hz) – f (175 Hz) c(131) – ab (208 Hz)	60 Hz – 200 Hz	110 dB	70 dB	10–18 ms
Bass drum		50 Hz – 400 Hz	115 dB	With pedal, 20 cm from drumhead: 120–128 dB / 80 dB	8–12 ms
Snare drum		80 Hz – 4 kHz	108 dB	55 dB	5–8 ms
Cymbal		100 Hz – 5 kHz	105 dB	65 dB	2–4 ms
Sizzling cymbal		100 Hz – 10 kHz	80 dB		
Triangle		2 kHz – 12 kHz			

Bibliography

Brixen, E. B. (1998). *Near field registration of the human voice: Spectral changes due to positions*. 104, Amsterdam, The Netherlands: AES Convention. Preprint 4728.

Dickreiter, M. (1989). *Tonmeister technology*. New York: Temmer Enterprices, Inc.

Fletcher, H. (1953). *Speech and Hearing in Communication*. New York: Van Nostrand.

Olson, H. F. (1952). *Musical Engineering*. New York: McGraw-Hill Book Company, Inc.

Pawera, N. (1981). *Microphones*. Dachau: Arsis Baedeker & Langs Verlag GmbH.

Traunmüller, H. and Eriksson, A. (1995). *The frequency range of the voice fundamental in speech of male and female adults. Manuscript*. Stockholms Universitet.

How Large
Is an Audio Signal?

CHAPTER OUTLINE

When describing the size of an audio signal, it is very important that there is agreement on what techniques to use to make the measurement. Otherwise there is a risk that differently measured values are not comparable.

ACOUSTIC SIGNALS

As a rule acoustic signals must be measured using transducers, for example microphones. An electrical signal that is analogous to an acoustic signal is what is really being used in practice when one performs this magnitude measurement.

ELECTRICAL SIGNALS

An electrical signal can be described by amplitude values for the voltage or current. Alternately, the electrical signal can be described by the energy it contains, for example the electrical power that is fed into a given load or during a given period of time.

When we study the amplitude of a signal as seen on an oscilloscope or a digital audio workstation (DAW), we can describe the values that we normally use to describe the magnitude of an audio signal.

PEAK VALUES

The peak value describes the instantaneous maximum amplitude value within one period of the signal concerned. The peak value can also be the maximum value that is ascertained during any period of time under consideration, even

Audio Metering. DOI: 10.1016/B978-0-240-81467-4.10005-X

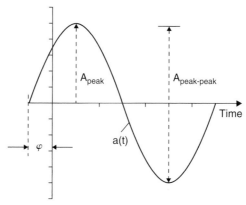

FIGURE 5.1 Sinewave where the peak value (U_p), peak-to-peak value (U_{p-p}) and the phase (φ) are indicated.

though the normal designation here concerns the maximum value within one period. Normally, the peak value measurement is used with symmetric signals, i.e., signals that deviate equally far from 0 in both a positive and negative direction.

For a sinusoidal wave:

$$
\begin{aligned}
\text{Peak value} &= \text{Maximum value} \\
&= \sqrt{2} \ \text{RMS value} \\
&= \frac{\pi}{2} \ \text{Average value}
\end{aligned}
$$

(See the following two sections for discussions of RMS value and average value.) In practical sound technology, particularly in so far as it concerns circuits for the recording and transmission of audio signals, it is the peak-to-peak value that is significant. It is this value that says how much "space" (voltage) the signal "takes up" in the electrical circuit, or how much a membrane must move in a purely physical sense from one extreme to the other in order to record or reproduce the sound.

For a sinusoidal wave:

$$
\begin{aligned}
\text{Peak to Peak value} &= \ 2 \times \text{Maximum value} \\
&= \ 2.828 \times \text{RMS value}
\end{aligned}
$$

In relation to the recording of audio signals, it is important to know that possible phase shifts, which can arise for example in different forms of pre-emphasis or correction networks, can make a signal asymmetric. (See Figure 5.2.)

In addition, there are many acoustic signals that contain an asymmetric wave. This applies, for example, to voice and to musical instruments, particularly percussion instruments, where the impact can be highly asymmetric.

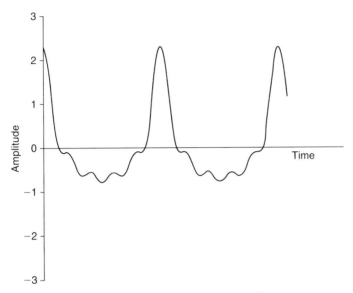

FIGURE 5.2 Example of a signal that is asymmetric around 0. The signal contains a fundamental tone and subsequently four harmonic overtones (symmetric sinusoidal tones). Each of the harmonics has a 90° phase shift in relation to the fundamental frequency. The length of this curve corresponds to two periods of the fundamental frequency.

AVERAGE VALUE

The average value is based on the average of the numerical values for the amplitude over a period of time. (Numerical value here means the indicated value without regard to its sign.)

The average value is calculated according to the following expression:

$$A_{Average} = \frac{1}{T} \int_{0}^{T} |\ a\ |\ dt$$

where
$T = $ period
$a = $ amplitude

For a sinusoidal wave:

$$\text{Average value} = \frac{2 \times \text{Maximum value}}{\pi}$$
$$= 0.9 \times \text{RMS value}$$

RMS VALUE

The RMS value is based on the energy that is contained in a given signal. RMS stands for root mean square and in essence is the square root of the square of the average value.

$$A_{RMS} = \sqrt{\frac{1}{T} \int_0^T a^2(t)\ dt}$$

where

T = period

a = amplitude as a function of time

For a sinusoidal wave:

$$\text{RMS value} = \frac{\text{Maximum value}}{\sqrt{2}}$$
$$= 1.11 \times \text{Average value}$$
$$= 0.707 \times \text{Maximum value}$$

For a square waveform:

$$\text{RMS value} = \text{Maximum value}$$

For a triangular wave:

$$\text{RMS value} = \frac{\text{Maximum value}}{\sqrt{3}}$$
$$= 0.576 \times \text{Maximum value}$$

For a half-wave:

$$\text{RMS value} = \frac{\text{Maximum value}}{2 \times \sqrt{2}}$$
$$= 0.354 \times \text{Maximum value}$$

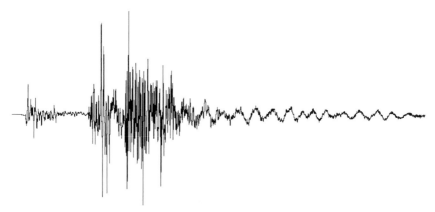

FIGURE 5.3 Sound impulse with a high peak value, but a low RMS value.

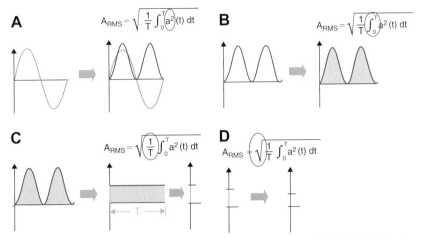

FIGURE 5.4 This strip explains what happens to the signal when finding its RMS value. A: All values are squared (provides positive figures). B: By integration the area below the curve is found. C: The area is divided by the time of the period. D: The square root is taken and we have a result: RMS!

Crest Factor

The crest factor expresses the relation between the peak value (maximum value) and the RMS value. It is a magnitude that is important to know when recording because the maximum value is an expression of the signal's amplitude, whereas the RMS value relates to what the level meter is showing.

$$\text{Crest factor} = \frac{\text{Maximum value}}{\text{RMS value}}$$

For a sinusoidal waveform:

$$\text{Crest factor} = \sqrt{2} = 1.414$$

For a square waveform:

$$\text{Crest factor} = 1$$

For a triangular waveform:

$$\text{Crest factor} = \sqrt{3} = 1.73$$

For pink noise:

$$\text{Crest factor} \approx 4$$

For (uncompressed) speech:

$$\text{Crest factor} \approx 10$$

Form Factor

The form factor expresses the relation between the signal's RMS value and average value. The form factor relates to the signal's waveform in that

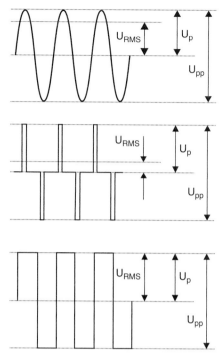

FIGURE 5.5 The relation between peak, peak-to-peak and RMS-values for different waveforms.

a small form factor is an indication of a flat waveform and a large form factor is an indication of a waveform containing peaks.

$$\text{Form factor}[\xi] = \frac{\text{RMS value}}{\text{Average value}}$$

For a sinusoidal waveform:

$$\text{Form factor} = \frac{\pi}{2\sqrt{2}} = 1.11$$

For a square waveform:

$$\text{Form factor} = \frac{2}{\sqrt{3}} = 1.15$$

The dB Concept

CHAPTER OUTLINE

When sound levels have to be specified, it is practical to do so in a manner that corresponds to the way in which the ear perceives them. Human hearing is approximately logarithmic with respect to the perception of both level and frequency. Logarithmic growth means that there is a constant proportion between the individual steps on the scale. For example, if 1 W is first imparted to a loudspeaker and then followed by 2 W, this will be experienced as a given step in perceived level. In order to experience an equivalent change of level again, not 3 W, but, rather, 4 W must be added — and 8 W for the next step. Here, there is a ratio of 2 between the individual steps.

In the calculation of dB, the base 10 logarithm is used. The base 10 logarithm of a number is that number which 10 must be raised to in order to give the number itself. Hence the logarithm of $100 = 2$, because $10^2 = 100$. Originally, the logarithm of the ratio between two power measurements was used (the Bel unit). Later, a tenth of this became preferred (the deciBel unit, abbreviated as dB).

The advantages of the dB unit are that large ratios can be described with the use of a maximum of 3 digits, and changes in a sequence will be the same throughout, expressed in terms of dB, regardless of whether the unit of measurement concerned is the volt, ampere, watt, and so on.

We will first look at the dB value for the ratio between the measurements of power. Power is measured in watts, so this transformation will apply to everything that can be measured in watts, in the form of acoustic as well as electrical

Audio Metering. DOI: 10.1016/B978-0-240-81467-4.10006-1

power (this transformation will thus not apply to amplitude values such as sound pressure, current, or voltage).

dB—POWER RATIO

The ratio between two measurements of power is expressed in the following manner:

$$10 \cdot \log \frac{P_1}{P_0}$$

where
P_0 = the reference (i.e., the value against which the comparison is being made)
P_1 = the relevant value to be specified

Example: I have a new amplifier. The old one was rated at 100 W, and the new one at 200 W. What is the difference in dB?

$$10 \cdot log\ (200/100)$$
$$\Downarrow$$
$$10 \cdot log\ 2$$
$$\Downarrow$$
$$10 \cdot 0.3$$
$$\Downarrow$$
$$3\ dB$$

The answer is that there is 3 dB more power in the new amplifier as compared to the old one.

dB—AMPLITUDE RATIO

The ratio between two amplitude values (for example sound pressure, current, or voltage) is expressed in the following manner:

$$10 \cdot \log \frac{(a_1)^2}{(a_0)^2}$$

where
a_0 = the reference (i.e., the value to be compared against)
a_1 = the relevant value to be specified

This expression can be reduced as the squaring is changed to multiplication when the logarithm is extracted:

$$20 \cdot \log \frac{a_1}{a_0}$$

Example: The input sensitivity of my amplifier is 0.775 V. The signal out of my mixer is only 0.3 V. What is the ratio expressed in dB?

$$20 \cdot log\ (0.3\ /0.775)$$
$$\Downarrow$$
$$20 \cdot log\ 0.39$$
$$\Downarrow$$
$$20 \cdot (-0.41)$$
$$\Downarrow$$
$$-8.2\ dB$$

The signal is −8.2 dB in relation to the input sensitivity. In other words, the signal is 8.2 dB below the level needed for full modulation of the amplifier.

Note that the dB value is positive when the ratio is above 1 and negative when the ratio is less than 1.

FROM dB TO POWER OR AMPLITUDE RATIO

One can calculate backwards from dB to power measurement ratios as follows:

$$\textbf{\textit{Power ratio} = 10}^{(x\ [dB]\ /10)}$$

Similarly, calculate backwards from dB to amplitude ratios as follows:

$$\textbf{\textit{Amplitude ratio} = 10}^{(x\ [dB]/20)}$$

Example: The sensitivity of a microphone in a catalog is stated as −56 dB re 1 V. What does that correspond to in terms of voltage?

$$10^{(-56/20)}$$
$$\Downarrow$$
$$10^{(-2.8)}$$
$$\Downarrow$$
$$0.00158\ (or\ 1.58 \cdot 10^{-3})$$

The result must be multiplied by the reference, which is 1 V.
The answer is thus $1.58 \cdot 10^{-3} \cdot 1$ [V], which is $1.58 \cdot 10^{-3}$ V or 1.58 mV

CONVERSION TABLE

Table 6.1 shows the conversion between power measurement ratios and dB as well as amplitude ratios and dB.

REFERENCE VALUES

In acoustics and electronics, specific reference values are used as a basis for the specified ratio in dB. In certain cases, the reference is stated directly, for example in the specification of sound pressure level: 60 dB re 20 µPa. This means that the sound pressure is 60 dB over $20 \cdot 10^{-6}$ pascal, which equates to a sound pressure of 0.02 pascal.

Instead of specifying the reference in full, in some situations one or more letters can be appended after the "dB" to indicate the reference. Following are some examples of this usage:

dBµ: Reference relative to 1 µW (microwatt)

dBd: Antenna gain in relation to a half-wave dipole

dBFS: For digital equipment (reference to "Full Scale")

TABLE 6.1 Convertion Between Power Ratios and Amplitude Ratios

dB	Power ratio	Amplitude ratio	dB	Power ratio	Amplitude ratio
0	1	1	0	1	1
1	1.259	1.122	−1	0.794	0.891
2	1.585	1.259	−2	0.631	0.794
3	1.995	1.413	−3	0.501	0.708
4	2.512	1.585	−4	0.398	0.631
5	3.162	1.778	−5	0.316	0.562
6	3.981	1.995	−6	0.251	0.501
7	5.012	2.239	−7	0.200	0.447
8	6.310	2.512	−8	0.158	0.398
9	7.943	2.818	−9	0.126	0.355
10	10	3.162	−10	0.100	0.316
12	15.85	3.981	−12	0.063	0.251
14	25.12	5.012	−14	0.040	0.200
15	31.62	5.623	−15	0.032	0.178
20	100	10	−20	0.010	0.100
26	398.1	19.95	−26	0.003	0.050
30	1000	31.62	−30	0.001	0.032
40	10^4	100	−40	10^{-4}	10^{-2}
50	10^5	316.2	−50	10^{-5}	$3.162 \cdot 10^{-3}$
60	10^6	1000	−60	10^{-6}	10^{-3}
70	10^7	$3.162 \cdot 10^3$	−70	10^{-7}	$3.162 \cdot 10^{-4}$
80	10^8	10^4	−80	10^{-8}	10^{-4}
90	10^9	$3.162 \cdot 10^4$	−90	10^{-9}	$3.162 \cdot 10^{-5}$
100	10^{10}	10^5	−100	10^{-10}	10^{-5}
110	10^{11}	$3.162 \cdot 10^5$	−110	10^{-11}	$3.162 \cdot 10^{-6}$
120	10^{12}	10^6	−120	10^{-12}	10^{-6}

dBi: Antenna gain in relation to omnidirectional antenna

dBk: Reference relative to 1 kW (kilowatt)

dBm: Reference relative to 1 mW (milliwatt) imparted at 600 ohm (often used incorrectly for the reference relative to 0.775 V RMS)

dBm0: Absolute power level (re 1 mW/600 ohm) at a point of 0 relative level

dBm0p: Absolute psophometric (frequency-weighted) power level re 1 mW/600 ohm at a point of 0 relative level

dBm0ps: Absolute psophometric (frequency-weighted) power level re 1 mW/600 ohm for a point of 0 relative level for program transmission (sound)

dBm0s: Absolute power level re 1 mW/600 ohm for a point of 0 relative level for program transmission (sound)

dBp: Reference relative to 1 pW (picowatt)

dBq0ps: Absolute and weighted voltage level re 0.775 V for a point of 0 relative level for program transmission (sound).

dBq0s: Absolute and unweighted voltage level re 0.775 V for a point of 0 relative level for program transmission (sound).

dBr: Relative specification in relation to an arbitrary reference, which must then be stated

dBrn: Relative specification in relation to noise floor; used in telecommunications

dBrs: Relative specification in relation to sound program level

dB SPL: Sound Pressure Level, with the reference 20 μPa

dBTP: (True Peak) Reference relative to the true peak of a signal.

dBu: Reference relative to 0.775 V RMS (in general the effective voltage that must lie over a resistance of 600 ohm in order for a power of 1 mW to be imparted); dBu is often used incorrectly for the reference relative to 1 μW

dBu0: Absolute voltage level re 0.775 V for a point of 0 relative level

dBu0s: Absolute voltage level re 0.775 V for a point of 0 relative level for program transmission (sound)

dBuv: American, reference relative to 1 μV

dBuw: American, reference relative to 1 μW

dBv: Often used in American data sheets for dBu (see dBu)

dBV: Reference relative to 1 V (volt)

dBW: Reference relative to 1 W (watt)

OTHER RELATIVE UNITS

VU: Understood as Volume Unit; used in audio metering. The change of 1 VU is a change of 1 dB. 0 VU has different definitions; however +4 dBm is commonly used. (See Chapter 12 on Standard Volume Indicator.)

LU: Loudness Unit; used in audio metering. Expresses the loudness of the program based on the electrical signal using frequency weighting (RL2B weighting, also called K-weighting) and by combining the level of all active channels to one single measure. Changing the level by 1 LU is a change of 1 dB.

Phon: Loudness unit, acoustic measure; equal to the sound pressure in dB re 20 μPa of an equally loud 1 kHz tone.

Sone: Measure of subjective loudness. One sone corresponds to 40 phons. A doubling/halving of the sone value corresponds to 10 phons (for example 2 sones equate to 50 phons).

Neper (Np): Used by the telephone companies.

1 Np = (20 lg e) dB ~ 8.686 dB.

1 dB = (0.05 ln 10) Np ~ 0.1151 Np.

WEIGHTED MEASUREMENTS

dB(A), dB(B), dB(C), dB(D) and dB(Z): These designations do not refer to a specific level, but rather to the fact that the signal has been measured with an inserted frequency weighting filter.

ADDITION OF dB

In order to find the total magnitude of two acoustic sound levels specified in dB the starting point will normally be that the two sound sources are not correlated (i.e., either the two signals do not resemble each other or they were recorded in a diffuse sound field). In practice, it is only in the near field between two loudspeakers that are reproducing the same sound where this condition would not be fulfilled.

The total sound level is calculated according to the following expression:

$$L_{p1} + L_{p2} = 10 \cdot \log \left(10^{\frac{L_{p1}}{10}} + 10^{\frac{L_{p2}}{10}} \right)$$

Example
What is 84 dB + 90 dB?

$$10 \cdot \log \left(10^{\frac{84}{10}} + 10^{\frac{90}{10}} \right) [dB]$$

$$\Downarrow$$

$$10 \cdot \log \left(0.251 \cdot 10^9 + 10^9 \right) [dB]$$

$$\Downarrow$$

$$10 \cdot \log \left(1.251 \cdot 10^9 \right) [dB]$$

$$\Downarrow$$

$$10 \cdot 9.1 \, [dB]$$

$$\Downarrow$$

$$91 \, [dB]$$

Nomogram (Addition)

One can avoid the work of performing the calculation by using a nomogram instead.

The difference is found between two sound levels (from the above mentioned example: 90 dB − 84 dB = 6 dB). This number is located on the horizontal axis. Then the vertical line is followed to the curve, after which you move horizontally to the left where the value 1.0 dB can be read. This value is then added to the highest level, which was 90 dB. The result is then $(90 + 1.0) = 91.0$ dB.

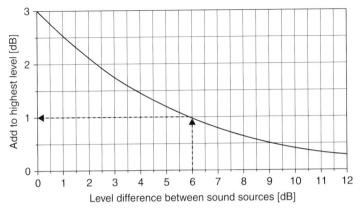

FIGURE 6.1   Nomogram for the addition of sound levels. The addition is performed on an energy basis and applies for uncorrelated sound sources (the sounds from the sources are different from each other) or sound sources in diffuse sound fields.

As the nomogram shows, two equally strong sound sources together would comprise a level that is 3 dB louder than the individual source. In practice, if the difference between the two sources is more than 10 dB then the weakest one can be ignored.

SUBTRACTION OF dB

Subtracting one sound level from another is related to the calculation of uncorrelated acoustic sound sources. This is calculated in the following way:

$$L_{p1} - L_{p2} = 10 \cdot \log\left(10^{\frac{L_{p1}}{10}} - 10^{\frac{L_{p2}}{10}}\right)$$

Example:
The noise level L_{p1} in a server room is 60 dB re 20 μPa. What is the level when one server having a noise level L_{p2} of 57 dB re 20 μPa is removed from the room?

$$10 \cdot \log\left(10^{\frac{60}{10}} - 10^{\frac{57}{10}}\right) [dB]$$

⇓

$$10 \cdot \log\left[\left(10^6 - 10^{5.7}\right)\right] [dB]$$

⇓

$$10 \cdot \log\left(0.5 \cdot 10^6\right) [dB]$$

⇓

$$57 [dB]$$

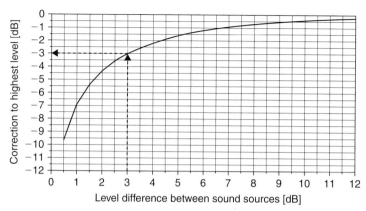

FIGURE 6.2 Nomogram for the subtraction of sound levels. The subtraction is performed on an energy basis and applies for uncorrelated sound sources (the sounds from the sources are different from each other) or sound sources in diffuse sound fields.

Thus the result is that the noise level is 57 dB re 20 μPa after removing one noise source.

Nomogram (Subtraction)

As with addition, subtraction can be performed using a nomogram.

The difference between the two sound levels is found (from the above mentioned example: 60 dB − 57 dB = 3 dB). This number is located on the horizontal axis. Then the vertical line is followed to the curve, after which you move horizontally to the left where the value −3 dB can be read. This (negative) value is then added to the highest level, which was 60 dB. The result is then (60 + (−3)) = 57 dB.

Notes

- A change of 1 dB is only just audible.
- A change of 3 dB is a clearly audible change.
- A change of 10 dB corresponds to a subjective doubling of the perceived sound pressure level.
- The addition of electrical and acoustic signals is covered later in Chapter 22 on summation of audio signals.

Chapter seven

The Ear, Hearing, and Level Perception

CHAPTER OUTLINE

Hearing is one of man's senses. Many would say the most important. How the sense of hearing actually works is still the subject of research. Only a brief review of the function of the ear will be given here. Selected important elements from psychoacoustics (the field that encompasses human perception of sound) will be presented.

Regarded in anatomical terms, the ear is divided up into three parts, namely the outer ear, middle ear, and inner ear.

THE OUTER EAR

The outer ear consists of the **pinna**, the **auditory canal**, and the **eardrum**. The pinna is of particular significance for directional perception (i.e., determining whether a sound is coming from the front or the rear). The auditory canal has dimensions that cause it to work as an acoustic resonator, which has an influence on the ear's frequency response. The eardrum has a diameter of approximately 9 mm (0.35 in.) and a thickness of approximately 0.1 mm (0.004 in.).

Audio Metering. DOI: 10.1016/B978-0-240-81467-4.10007-3

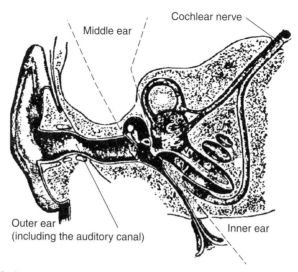

FIGURE 7.1 A cross-section of the human ear is shown here. The outer ear with the pinna and auditory canal. The eardrum comprises the transition to the middle ear with the three ossicles (the hammer, anvil, and stirrup), which via the oval window are connected to the inner ear that contains the cochlea.

THE MIDDLE EAR

Three small bones, called **ossicles**, are contained in the middle ear. These are named the **hammer, anvil**, and **stirrup**, respectively. These three bones function as a system of levers that ensure that the sound pressure variations in the outer ear are transferred to the inner ear. The system could be called a pressure transformer in that the pressure is increased by approximately a factor of 15 between the eardrum and the inner ear.

The hammer, which sits on the eardrum, is approximately 8 mm (0.32 in.) long and weighs approximately 25 mg (0.0009 ounce). The stirrup, which is connected to the inner ear, is approximately 3 mm (0.12 in.) and weighs approximately 3 mg (0.0001 ounce).

The middle ear is filled with air and is connected to the nasal cavity via the **Eustachian tube**, which is a narrow tube that opens when a person yawns or swallows. This allows for equalization of any possible pressure differences between the surrounding environment and the middle ear.

THE INNER EAR

The inner ear is a fluid-filled organ that is well-protected with its placement in the cranial tissue. The inner ear is an organ that never grows and for which there is no form of regeneration of destroyed cells.

The inner ear has a snail-shaped cavity that is called the **cochlea**. With its slightly more than 2.5 turns, the cochlea has a volume of approximately 100 mm^3. The cavity is divided into two parts by the **basilar membrane**, save for a small hole (the **helicotrema**) at the top of the cochlea.

The width of the basilar membrane varies from approximately 0.1 mm (0.0039 in.) to 0.5 mm (0.02 in.). The footplate of the stirrup is connected to the **oval window**, one of the cochlea's two openings. The **round window**, which is the other opening, ensures pressure equalization when pressure is exerted on the oval window.

The **organ of Corti** sits on the basilar membrane and is equipped with thousands of hair cells placed in rows. A gelatinous membrane, the **tectorial membrane**, sits above these rows. When the ear is affected by a sound wave, the stirrup's movements will be converted to a movement of the fluid in the inner ear. This sets the basilar membrane in motion, which in turn causes a shearing motion between the organ of Corti and the tectorial membrane. The tectorial membrane activates the hair cells, releasing electrical impulses that are sent to the brain via the **auditory nerve**, which consists of approximately 25,000 nerve fibers. The upper cut-off frequency for the transmission of signals to the brain is approximately 1 kHz.

SENSITIVITY OF THE EAR

The sensitivity of the ear varies with the frequency. Due to the shape of the auditory canal, the sensitivity is greatest in the range around 2-4 kHz.

This frequency dependence is not constant, but rather changes with the loudness of the sound. This is described in more detail later. By and large, it is possible to say that the ability to hear bass increases with the sound pressure. A person will thus perceive differences in the sound of an acoustic image depending upon the sound pressure level it is reproduced at. Accordingly, it is always important for audio engineers to perform a listening test of program material at the level for which later use is being contemplated.

HEARING LOSS

Age-related hearing loss affects the highest frequencies in the audible spectrum and begins at around 25−30 years of age.

Noise-related hearing loss can arise from individual events such as explosions and screeching loudspeaker systems. However, the long-term influence of larger doses of noise can also lead to permanent hearing loss, like extensive use of loud-playing earphones. In these cases, the hair cells in the inner ear fall over or become deformed, and no regeneration takes place. Before the actual noise-related hearing loss sets in, it can be ascertained that the ear's ability to differentiate between frequencies has deteriorated.

Tinnitus is a particularly annoying form of hearing disorder. In its mild form, which many musicians are familiar with, a ring or a hiss arises in the ears as a consequence of the influence of loud sounds. This will normally subside after a few hours; however, it may never disappear. A person with tinnitus will always hear a sound that can be perceived as a ringing, hissing, screeching, or roaring. This is sufficient reason to take care of your hearing.

THE EAR AS A FREQUENCY ANALYZER

Under the influence of sound, the basilar membrane undergoes fluctuations of a certain size. These fluctuations are approx. 10^{-10} mm ($4 \cdot 10^{-12}$ in.) at the threshold of hearing. Depending on the frequency that is affecting the ear, the fluctuation will have a maximum at a certain distance from the oval window. The position of this maximum is significant with regard to the ear's ability to determine the frequency of the sound, since the nerve fibers of the organ of Corti also function as selective filters with a certain bandwidth. For tones in the middle frequency range above 200 Hz, the ear can differentiate frequency variations of less than 1%.

When the ear has to differentiate between the frequency components in complex sounds, the filter function works as a series of parallel filters, each with a certain bandwidth. Until recently, it had been supposed that these frequency bands, which are referred to as "critical bands," had a bandwidth of approximately 20% of the center frequency (almost comparable with the bandwidth in a 1/3-octave filter). More recent research has shown, however, that this bandwidth is closer to 10%.

LOUDNESS AS A FUNCTION OF FREQUENCY

As described previously, human hearing is frequency-dependent. Moreover, this frequency dependence varies with loudness. The extent of this relationship is expressed in the "equal loudness" curves. These curves were created through the testing of a large number of people with normal hearing.

The basis is pure tones in a free, frontal field; which is to say that the test subjects in practice have found themselves in an anechoic chamber with a loudspeaker in front of them. The threshold value is then determined for the subject, i.e., the sound pressure precisely sufficient for a tone at a given frequency to be heard. In addition, as a function of frequency, the required sound pressure for constant loudness is also registered. This is done by comparing the tone concerned with a tone at 1000 Hz and at a known sound pressure level. The curves thus express constant loudness — all points on a curve sound equally loud. The unit for loudness is the phon.

Example:

The loudness for a pure tone with a frequency of 1 kHz and a sound pressure of 50 dB re 20 µPa is 50 phons.

The loudness of a pure tone with a frequency of 100 Hz and a sound pressure of 50 dB re 20 µPa is approximately 40 phons.

LOUDNESS AS A FUNCTION OF THE SOUND FIELD

The direction of a sound is also significant to its loudness. A frequency dependency occurs again. This can be significant in the assessment of acoustic images in a loudspeaker arrangement for the reproduction of surround sound. The timbre can appear to be different depending on whether the sound comes from the center speaker, from the right or left front speaker or from the right or left rear speaker.

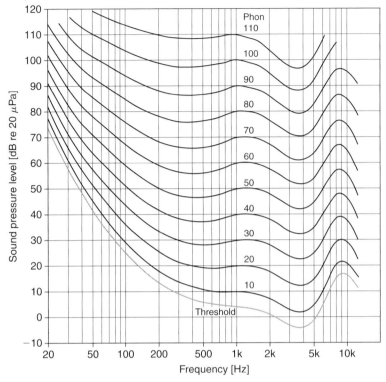

FIGURE 7.2 The equal loudness curves as they are standardized in ISO 3746. The curves express the sound pressure level at which pure tones must be represented within the free field in order to experience the same level of hearing regardless of the frequency. The lower curve describes the MAF (Minimum Audible Field) threshold value.

It is thus clear that the sound in a center speaker cannot be directly replaced by a phantom image created by the left and right front speakers. The relationship between a frontal direct sound field and a diffuse sound field, where all sound directions are equally probable, is shown on the curve below.

The lower curve describes the MAF (Minimum Audible Field) threshold value.

LOUDNESS AS A FUNCTION OF THE DURATION OF A SIGNAL

The duration of a signal is as important to the loudness of the signal as the frequency. This is one of the key questions when it comes to dynamic range: How long is the signal and how loud does it sound?

Studies have shown that signals (tones) with duration as low as approximately 200 ms sound just as loud as an equivalent constant signal at the same sound level. If the duration of the signal is less than 200 ms, the perceived loudness is reduced.

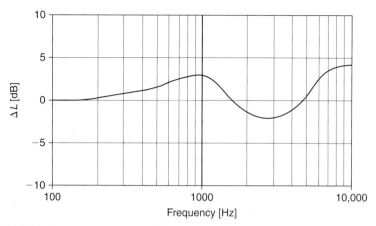

FIGURE 7.3 The curve shows the difference between the perception of a direct, frontal sound field and a diffuse sound field.

The curves in Figure 7.4 show the necessary increase in level in order to maintain constant loudness. For example, a signal of 10 ms duration must be increased by 10 dB in order to sound just as loud as an equivalent constant signal. Standardized instruments such as a peak program meter (PPM) are able to show the full level for signals with a duration of 10 ms. A PPM does not take into consideration how loud the signal sounds, but rather its physical magnitude. A VU meter has an integration time of 300 ms; an LU meter has a short-term integration time of 400 ms. Hence these instruments far better illustrate loudness, or volume, especially in so far as it concerns the duration of the signal. Another part of the assessment of the duration of a signal is the fact that the perceived loudness will fall during extended exposure due to the effects of fatigue.

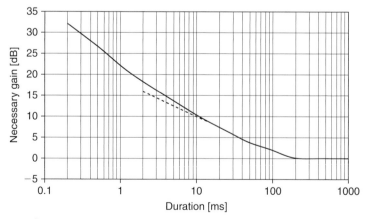

FIGURE 7.4 The curves describe the required increase in the level of impulses in order to maintain the same loudness as a constant signal. The fully connected line applies to broadband noise. The dotted line applies to pure tones.

MASKING

If the ear is subjected to a sound in a limited part of the frequency spectrum (for example a pure tone or a narrow band) then this sound, depending on the level, will to a certain extent mask or "hide" sounds of a similar or slightly higher frequency and lower level, even though this level also lies above the threshold values for the sensitivity of the ear. If the ear is subjected to two pure tones that are very close to each other in terms of frequency, then these will mask each other to a greater extent than if these tones were further apart from each other. Two tones that are close to each other will often be perceived as one tone that varies in its level (differential tones).

These phenomena have been studied through experiments that are similar to those that were carried out to ascertain the sensitivity of the ear, that is by the test subjects being asked to adjust the level of individual tones of different frequencies until they are just barely audible. This experiment differs simply in that the ear is constantly subjected to a fixed tone at 1 kHz with a sound pressure level of 80 dB. The results show that a tone of 1.5 kHz in this situation must have a level of approximately 45 dB above the threshold value in order just to be heard at the same time as the disrupting tone at 1 kHz. This phenomenon does not exist only at 1 kHz. Similar results can be obtained in the other part of the frequency spectrum, just as noise confined to a narrow bandwidth can have approximately the same effect as pure tones.

FORWARD AND BACKWARD MASKING

Masking does not occur only with simultaneous tones, but can also occur on the basis of sounds that start or stop split seconds before or after the masking or dominant sound.

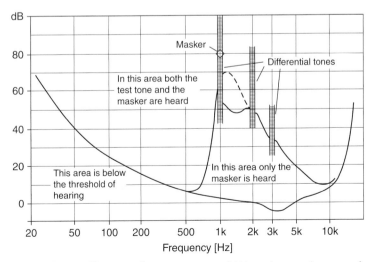

FIGURE 7.5 The masking curve for a pure tone at 1 kHz and a sound pressure level of 80 dB re 20 μPa.

Forward masking refers to the phenomenon whereby a tone can be masked by a sound that has stopped up to 20–30 ms before the tone concerned starts. It appears as if hair cells that have just been stimulated are not nearly as sensitive as cells that have rested for a longer period of time.

Backwards masking refers to a tone being able to be masked by a sound with a higher level that begins up to 10 ms after the tone has begun. This effect evidently does not originate in the inner ear, but rather in the brain. It appears that the brain's processing of the weaker tone is set aside when a sound impulse with a greater intensity appears.

The masking ability of the ear is utilized in connection with noise reduction and in particular with "intelligent" bit reduction. The algorithms used are based upon "perceptual coding."

AUDIBILITY OF PEAKS AND DIPS

When changes are made to the frequency response in an electro-acoustic system, tops, i.e., local rises in frequency response, are more audible than corresponding dips. Experiments show that a rise of just 10 dB in a narrow band around 3.2 kHz in an otherwise neutral frequency response would be perceived by all people with normal hearing. Correspondingly, a drop in the same frequency band of 10 dB is only perceived by approximately 10% of the listeners. A drop of 20 dB would only be perceived by approximately 40%.

This knowledge can be utilized in particular for the adjustment of loudspeaker systems where limiting the possibilities of acoustic feedback is desired for certain frequencies. It can also be used for removing disturbing tonal noise from program material. If for inexplicable reasons a person has accidentally recorded a test tone (pure tone) on top of the program material, the tone

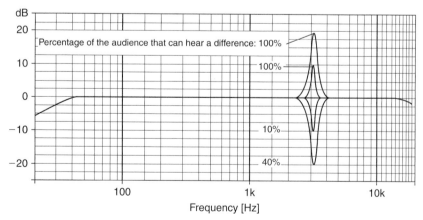

FIGURE 7.6 Based upon psychoacoustic experiments, it has been demonstrated that everyone can hear amplification in a narrow frequency range, whereas only 10% can hear a corresponding attenuation of 10 dB. If the drop is 20 dB, only 40% can hear the change.

can be removed with a notch filter, which has the property that it only attenuates a very limited frequency range. What would disappear in addition to the test tone would generally not have an audible effect on the program material.

THE LAW OF THE FIRST WAVE FRONT

The ability to localize a sound source in a closed space is affected by the reflections in the space. However, experience shows that even in a room with many reflections — and hence a long reverberation time — it is possible to localize a single sound source even if the sound energy of the direct sound is much smaller than the energy of the reverberant sound field at the position of the listener. This phenomenon is called the **law of the first wave front** and was first described as such by Lothar Cremer in 1948.

A study carried out by Helmut Haas in the 1950s has become important background information in the understanding of how the hearing system perceives the combination of direct and delayed sounds. In this experiment subjects were placed in an acoustically damped room in front of two loudspeakers (LS1 and LS2) in a stereo setup with a listening angle of 80 degrees. A speech signal (approximately 5 syllables per second) was reproduced by both LS1 and LS2, however with a possibility of changing both the level and the delay of the sound reproduced by LS2. The level of LS1 was approximately fixed at 50 dB re 20 µPa.

With the level and delay of LS2 changing, the subjects' task was to select one of three options: LS1 was inaudible, LS2 was inaudible, or the auditory event was localized in a direction right between the two loudspeakers LS1

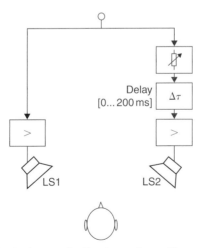

FIGURE 7.7 This is the basic setup for Haas' experiment. The same signal is reproduced by loudspeakers LS1 and LS2; however, the level and delay of LS2 can be controlled. By adding various delays to the LS2 signal, the task of the listener is to define what the level difference is between LS1 and LS2 when (1) Only LS1 is audible, (2) only LS2 is audible, and (3) when the sound is localized to a position right between LS1 and LS2.

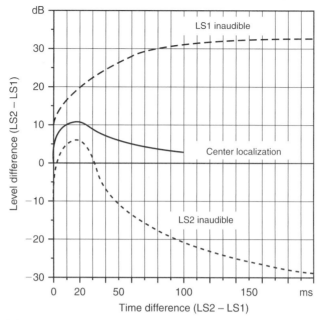

FIGURE 7.8 These curves represent the result of the experiment, the level difference between LS1 and LS2 (LS2 − LS1). The upper curve shows that at zero delay LS2 must be 10 dB louder than LS1 if LS1 must not be audible. However, by increasing the delay the level difference also has to be increased, with LS2 needing to be even louder. The middle curve shows that LS2 must be louder in order to keep center localization when delay is added. The bottom curve is the most interesting. When LS2 is delayed 15—20 ms the level of LS2 can be up to 5 dB louder than LS1 without being audible.

and LS2. In Figure 7.8 the main results are shown. In electro-acoustic applications the lower curve is the most interesting. It shows that if the sound from LS2 arrives approximately 5 and 35 ms later than the first arrived sound (from LS1), then the level of LS2 can be up to 5 dB louder than LS1 without being audible. This is often referred to as the "Haas effect."

If the delay exceeds 32—50 ms (depending on the type of signal reproduced) the delayed sound will be perceived as an echo. It is easier for the ear to detect echo effects on impulsive/percussive sounds than more sustained sounds.

Bibliography

Blauert, J. (1997). Spatial Hearing. The MIT Press.
Evans, E. F. (1993). Basic Physiology of the Hearing Mechanism. AES 12th international conference.
Juhl Petersen, O. (1974). Menneskets lydopfattelse. (Human Perception of Sound). Laboratoriet for Akustik, DtH: (Technical University of Denmark).

Time Weighting

CHAPTER OUTLINE

Most instruments that display levels create the value displayed through an RMS averaging of the signal over time. In practice, two different methods are used to create the averaged signal, namely linear and exponential averaging. The term **time weighting** refers to the time window that is used for the creation of the value displayed.

LINEAR AVERAGING

With linear (time) averaging, an average level is created as a simple RMS value of the signal within a given period of time, for example 2 seconds. The weighting for linear averaging is shown in Figure 8.1.

On instruments that are used for acoustic measurements of sound, there is typically an option with a selection of averaging times to choose from:

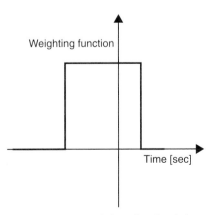

FIGURE 8.1 Principle of linear time weighting: the signal that occurs within the time window is averaged.

Audio Metering. DOI: 10.1016/B978-0-240-81467-4.10008-5

70 ms, 250 ms and 2 s corresponding to the terms I (impulse, not widely used anymore), F (fast), and S (slow), respectively.

For measurements with program level meters for audio recording and transmission, the averaging times used are typically 0.1 ms (fast), 10 ms (standard PPM), 300 ms (VU meter), and 400 ms (LU meter, short term).

EXPONENTIAL AVERAGING

With exponential averaging, the weighting is an exponential function, which, as opposed to linear averaging, does not extend over a fixed period of time. The RMS signal created will thus "remember the past," but in such a manner that events that lie far away in the past will have a lesser weight than events that have just occurred.

Exponential weighting is shown in Figure 8.2. With exponential averaging, the concept of a time constant is used rather than a time period for averaging. The time constant is a measure of how fast the exponential function "dies" out. Or more precisely, it specifies the time before the exponential function is reduced to 69% of its beginning value.

In connection with the acoustic measurement of sound, there normally is an option to select time constants of 125 ms (fast) and 1 s (slow), as well as an option for impulse weighting.

IMPULSE

Impulse weighting differentiates itself from fast and slow weighting by the fact that it contains a peak detector. It is thus the peak value rather than the RMS value that is decisive for the impulse value displayed when the instrument or meter is in I, or impulse, mode. As the name suggests, this form of weighting

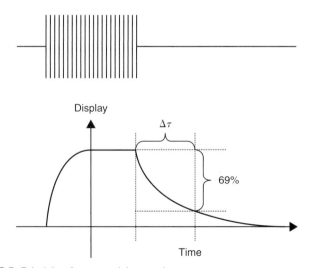

FIGURE 8.2 Principle of exponential averaging.

is intended for the measurement of impulse noise. The time constant is small (35 ms), and a peak detector with an ensuing long time constant (1.5 s) will ensure that impulses will take a long time to "die" out. This form of level display is used, among other things, for the measurement of noise in the external environment, for example when measuring noise from shooting ranges. However, it must be mentioned that the term "I" is gradually finding its way out of the standards. Instead most standards now refer to "peak" measurements, which are displayed with a hold function that leaves the indicator at the max peak detected during the measurement. Depending on national legislation this value is measured using A- or C-frequency weighting.

PEAK

In many countries the measurement of impulse "I" at the workplace has been superseded by the measurement of the peak level. In this case a frequency network (A- or C-weighting) may or may not be used. The peak measurement is supported by a hold function that maintains the highest peak level measured until the next reset.

THE TIME FACTOR IN PROGRAM LEVEL METERS

In program level meters, time weighting and integration time are defined in a slightly different manner as compared to equipment for acoustic measurements.

The integration time is defined in the standards as the time it takes for a signal to reach close to full amplitude. Or, to be more precise, the IEC standard (IEC 60268-10) as an example states that the integration time is the time it takes a 5 kHz sinusoidal burst at reference level to reach a display 2 dB under that reference level. There must be a silent interval between each tone burst large enough that the instrument can settle back to its minimum.

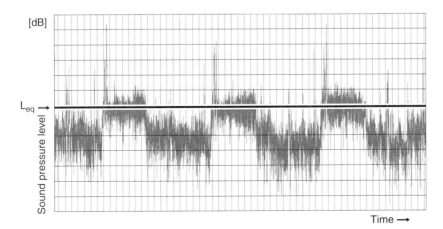

FIGURE 8.3 L_{eq} of a time-varying sound pressure level.

In order to be able to see what the level was when it was there, a fallback time can also be defined. This is the time that elapses from a constant signal being broken until the display reaches a specific lower point on the scale. This can typically be specified in dB/s.

EQUIVALENT LEVEL, L_{eq}

L_{eq} expresses the energy equivalent level. The L_{eq} value is an average value of (on an energy basis) the level over a longer interval of time (minutes to hours). This concept is used in connection with acoustic measurements of sound; however, L_{eq} is also used in connection with electrical measurements for calculation of the loudness of program material. Normally the value is calculated from a frequency-weighted signal, e.g., IEC A (acoustical measurements) or ITU K-weighting (loudness, long term, program material):

$$L_{eq}(w) = 10 \log_{10}\left(\frac{1}{T}\int_{t_1}^{t_2} a_w^2(t)dt\right)$$

where
$\qquad$ w = frequency weighting, e.g., IEC A or ITU k
$\qquad a_w(t)$ = amplitude of the weighted curve signal
$\qquad t_2 - t_1$ = integration time

Chapter | nine

Frequency Weighting
and Filters

CHAPTER OUTLINE

Audio Metering. DOI: 10.1016/B978-0-240-81467-4.10009-7
Copyright © 2011 Eddy B. Brixen. Published by Elsevier Inc. All rights reserved.

Frequency weighting involves emphasizing/suppressing certain parts of the spectrum of an audio signal. There are two reasons for using frequency weighting:

- To take the human perception of sound into account when measuring and specifying sound levels (acoustic) or the internal noise of devices and systems (electrical).
- For the pre-emphasis of audio signals, to obtain an optimization of the frequency response or signal-to-noise ratio.

WEIGHTING EMULATING THE RESPONSE OF THE EAR

Frequency weighting that responds approximately like the ear's hearing ability is utilized in particular when noise is measured. It can involve acoustic noise or electrically generated internal noise in devices or systems. In all cases, it is the audibility of the noise that one wants to measure. However, program material for broadcast is now also measured and assessed in relation to human hearing.

Typically the generally used weighting filters suppress lower frequencies and slightly emphasize the range just above 1 kHz. This means that the low frequencies do not have as large an influence on the result of the measurement as do those frequencies that lie between say 1 and 4 kHz.

In practice, the weighting filter concerned is inserted into the electrical part of the measurement sequence. However as a rule, the filter will be built into the measuring device concerned, such as a sound pressure meter or an audio analyzer.

IEC A

A-weighting is used in particular for acoustic measurements of noise. However, it is also rarely used in connection with electrical measurements. A-weighting is specified in IEC 60 651, but references are also made in other standards.

Sound measurements performed with A-filters are often described with an "A" in parentheses, for example xx dB(A) or with an "A" as an index, $L_A = $ xx dB.

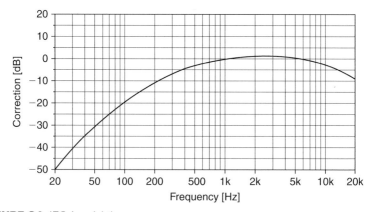

FIGURE 9.1 IEC A weighting curve.

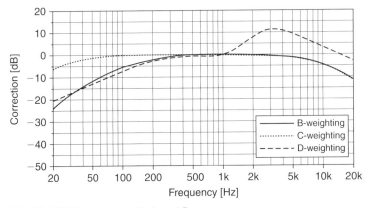

FIGURE 9.2 Weighting curves B, C, and D.

IEC B, C, and D

The B, C, and D weightings ultimately belong together with A-weighting, as the idea was that different weightings should be used depending of the level of sound concerned. A-weighting should be used at the weakest sound levels, since the ear does not hear that much bass at these levels. At higher levels a change should be made first to the B- and later to C-weighting.

Today, the original B-weighting standard is by and large not used at all. However, a part of it has been implemented for loudness measures of program material.

C-weighting is used in the measurement of the maximum sound pressure levels of PA systems and cinema loudspeaker arrays, and also in some cases for the measurement of monitoring levels. D-weighting is used in connection with the measurement of noise from aircraft.

IEC Z

Z-weighting (IEC 61.672-1) is for use with acoustic measurements and is actually no weighting (Z for zero-weighting). However, it defines a flat frequency response of 10Hz to 20 kHz $\pm$ 1.5dB. This response replaces the older "linear" or "unweighted" responses as these did not define the frequency range over which the meter would be linear.

RLB — Revised Low-frequency B-weighting (ITU-R BS.1770)

Resent work by the ITU has led to the conclusion that the loudness of a program can be determined by a relatively simple algorithm involving a weighting curve that is basically a high-pass filter with approximately the properties of the low frequency part of the IEC B-weighting curve. Hence the name: RLB — Revised Low-frequency B-weighting.

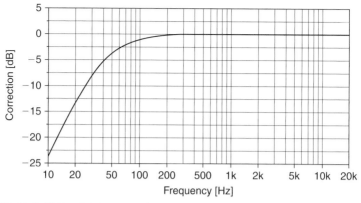

FIGURE 9.3 RLB weighting curve (ITU 1770)

K-weighting

K-weighting is a further development of the RLB curve. Early on this curve was called RL2B, but now it is known as K-weighting. This curve includes a high-frequency shelving curve to compensate for the presence of the head. More about this curve is found in Chapter 14 on Loudness Metering.

ITU-R BS.468-1 (CCIR)

This weighting curve (known in the past as CCIR 468) is used in the measurement of noise in electro-acoustic devices, predominantly broadcast equipment. As opposed to most measuring systems, RMS is not used, but rather a quasi-peak detector (here, quasi means "the same as").

CCIR/ARM

In 1979, Dolby Laboratories presented a simplified procedure whereby the signal-to-noise ratio in electro-acoustic equipment could be measured. The

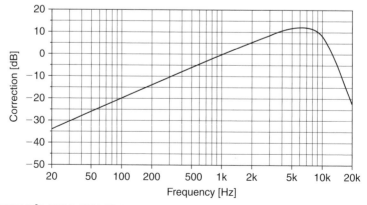

FIGURE 9.4 ITU-R BS.468-1 weighting curve (formerly "CCIR").

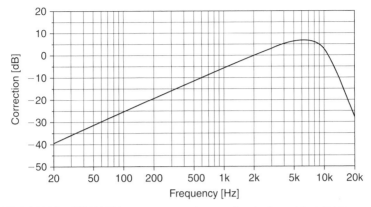

FIGURE 9.5 The CCIR/ARM curve crosses the zero-line (unity gain) at 2 kHz.

desire was for a weighted measurement that could be performed with relatively inexpensive equipment. In addition, it had to provide a value that was easily understandable and which could be "swallowed" by the industry in general.

The result became "CCIR/ARM," which uses CCIR weighting (now ITU-R BS.468), but with a 5.6 dB offset, making the curve cross 0 dB (unity gain) at 2 kHz instead of 1 kHz. The value is read using a simple millivoltmeter that shows the average value. "ARM" stands for Average Responding Meter.

This weighting has subsequently been used in Dolby's system for measuring sound levels in cinema films. In this case, the average is not used, but, rather, an approximated L_{eq} (equivalent level). The result is specified as $L_{eq}(m)$, where the "m" stands for movie.

Other Weightings

Other frequency weightings exist, of which some are more specialized and some are losing popularity. The first-mentioned group includes, for example, psophometer curves, which are used for frequency weighting especially in the measurement of telephones and telephone components. Others include weightings for the measurement of the capacity of loudspeakers, since the assumption is that not all frequencies are represented at the same level in the average music signal.

Among the soon-to-be-forgotten weighting curves are those used in the measurement of rumble and the like in gramophones (those things used to play vinyl records).

EMPHASIS

Pre-emphasis and de-emphasis are used in a number of contexts, particularly where analog signals have to be recorded or transmitted. The reason for the use of emphasis is to attain a good frequency response or to attain a good signal-to-noise ratio, or a combination of both.

On transmission lines this typically involves solely the raising of the treble. A large amount of attention must be devoted to the measurement of the signal output for an FM transmitter or for satellite equipment, as that output must be measured with pre-emphasis included, even if a limiter is on the output. With pre-emphasis, there is risk of burning out a transmitter or knocking out the satellite link due to overamplification. Even though a limiter has been placed on the output, it cannot save the situation if it is not connected *after* the pre-emphasis.

With digital signals, it is also possible to get mixed up in something that can later be quite difficult to solve. The AES3 standard (formerly AES/EBU) allows for the use of different pre-emphasis types. If signals are combined in a mixer where some of the signals are emphasized and others are not, it can result in mixed signals that subsequently cannot be processed properly. Hence it is a cardinal rule that pre-emphasis is avoided on signals that are to be mixed.

The μs Concept

The μs concept refers to the time constant that an RC element must have in order to attain the relevant frequency characteristic.

The time constant can be expressed as follows:

$$\tau = R \cdot C$$

where
τ = time constant [s]
R = resistance [Ω]
C = capacitance [F]

or

$$\tau = 1/2 \cdot \pi \cdot f_0$$

where:
τ = time constant [s]
f_0 = corner frequency of the emphasis circuit (the 3 dB point).

This is a first-order filter, i.e., the slope of the inclined part of the curve is 6 dB/octave.

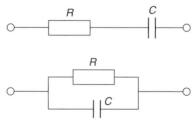

FIGURE 9.6 RC circuits.

TABLE 9.1 Time Constants of 1st Order Rc-Circuits and the Corresponding Corner Frequencies.

Time constant [µs]	Frequency [Hz]
17.5	9094.57 ≈ 9100
35	4547.57 ≈ 4550
50	3183.10 ≈ 3180
70	2273.64 ≈ 2275
75	2122.01 ≈ 2120
90	1768.39 ≈ 1770
120	1326.29 ≈ 1325
318	505.00 ≈ 500
3180	50.50 ≈ 50

A number of standardized time constant-based pre-emphasis techniques are used, some in transmission techniques and others in tape recording. Some of the most frequently used time constants are listed here along with the corresponding corner frequencies:

50/75 µs

50 µs or 75 µs is used in particular for FM transmission, either involving a normal radio broadcast or transmission systems for wireless microphones.

The advantage is an improved signal-to-noise ratio, especially at higher frequencies. The drawback is that the curve rises endlessly towards higher frequencies. One then becomes dependent on frequency-limiting a signal during FM broadcasts among other things. In Europe, broadcasters generally employ a 50-µs time constant for FM transmission whereas American stations use 75 µs.

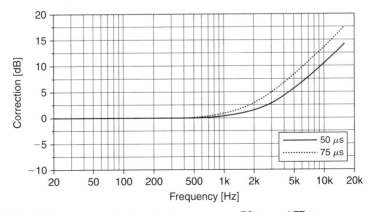

FIGURE 9.7 Emphasis defined by the time constants 50 µs and 75 µs.

TABLE 9.2 Emphasis for Tape Recording. The 3180μs correction is Introduced to Raise the Signal Above Hum and the Short Time Constants Have the Purpose of Compensating for Losses.

	Standard	Tape speed [cm/2]	Time constants [μs]	
Reel-to-reel	NAB	38	3180	50
	NAB	19	3180	50
	NAB	9.5	3180	90
	NAB	4.75	3180	90
	IEC 1	76	–	35
	IEC 2	76	–	17.5
	IEC 1	38	–	35
	IEC 2	38	3180	50
	IEC 1	19	–	70
	IEC 2	19	–	50
	IEC	9.5	3180	90
Cassette tape	IEC I	4.75	3180	120
	IEC II	4.75	3180	70
	IEC III	4.75	3180	70
	IEC IV	4.75	3180	70

50/15 μs

The pre-emphasis curve employed by consumer digital formats such as CD and DAT rises at high frequencies in the same way as the 50 μs curve for FM radio, but due to the second time constant (corner frequency @ 10.6 kHz) the gain levels off towards the top of the audio band, reaching around 10 dB by 20 kHz.

Emphasis in Tape Recorders

In tape recorders, different time constants are used in the shaping networks of the recording and playback circuits. Even though analog tape recorders are on the way out, it is still essential to keep track of the time constants for handling archived material. Mixing the standards can displace the tonal balance of a recording significantly.

Two time constants are used in certain standards, since correction is performed at both ends of the spectrum. The correction at low frequencies (3180 μs) is intended to lift the signal up over hum components.

Here are some of the standards:

J.17

J.17 emphasis is named after the ITU standard in which its description is found. It is a weighting that is used in connection with the pre-emphasis of signals that

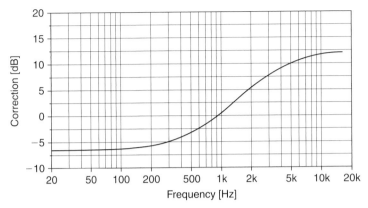

FIGURE 9.8 ITU J.17 emphasis.

are either transmitted or stored. This pre-emphasis is used for NICAM stereo sound for TV and the audio channels of satellite transmissions.

J.17 pre-emphasis not only increases high frequencies, but also reduces low frequencies, both around 10 dB relative to its unity gain at 2 kHz.

The advantage of this pre-emphasis in comparison with 50 μs or 75 μs is that the rise flattens out towards higher frequencies. The level is thus less sensitive to the selection of the upper frequency band.

RIAA

RIAA is an abbreviation for the Recording Industry Association of America. The RIAA emphasis is used in the recording of gramophone records. The purpose is in part to give a correct frequency response and in part to "economize" the needle's movement in the groove.

If one were to cut a record with the same amplitude at all frequencies, the physical variation of the groove would be greater the lower the frequency that is being recorded. The pickup would have a difficult time tracking, and there would be less room for music on the record. Therefore, the recording curve falls at a slope of 6 dB/octave towards the lower frequencies from a transition frequency of 500 Hz (318 μs). However, the curve flattens out below 50 Hz (3180 μs). Above a transition frequency of approximately 2.1 kHz (75 μs), the recording curve rises by 6 dB/octave. A constant amplitude could well have been chosen for higher frequencies, but the choice was made to introduce this rise. The reasons include avoiding dust particles and the like, which have a disproportionately large and destructive effect on the acoustic image. In practice, it is difficult to see on the curve that the interval between 500 Hz and 2 kHz has constant amplitude.

In 1976 IEC introduced a modification of this curve by introducing a new time constant that modifies only the extreme bass. It is known as the RIAA/IEC curve. This correction has never achieved big success; the original RIAA curve is still the most widely used.

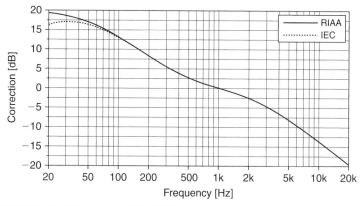

FIGURE 9.9 The RIAA and the RIAA/IEC curves for gramophone records (with a minor correction in the low bass range in the RIAA/IEC curve).

If working with archives it is recommended to look up the countless number of special equalization curves that have been introduced for specific recording systems invented and implemented through more than hundred years of disc recording.

FILTERS IN AUDIO

Electrical filters are used everywhere in audio. As a rule, they have an effect on the level of the signal. In addition to the concepts already mentioned, some of the fundamental general definitions will be described here.

Cutoff Frequency

This is the frequency at which a filter begins to attenuate. It is often defined by the attenuation being 3 dB at the frequency concerned.

Bandwidth

This is the distance between the −3 dB cutoff points on a frequency response curve, i.e., $f_{upper} - f_{lower}$. Expressed either in absolute terms in Hz or in relative terms in octaves (often in 1/1 octave, 1/3 octave or in fractions of an octave expressed with decimals, for example 0.1 octave) or in percentages:

$$b = \frac{(f_u - f_l) \times 100}{f_c} \, [\%]$$

where
f_u = upper cutoff frequency
f_l = lower cutoff frequency
f_c = center frequency
1/1 octave ~ 70%, 1/3 octave ~ 22%. See also Q-Factor described below.

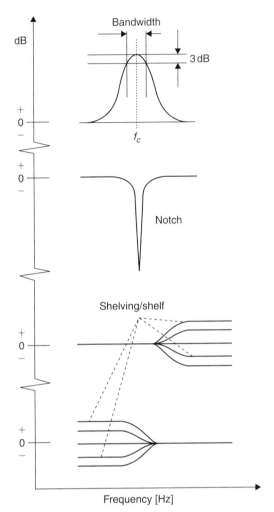

FIGURE 9.10 Upper curve: Band-pass filter. The bandwidth is defined (in Hz) by the points on the curve where the attenuation is 3 dB. Middle curve: The notch filter is a "narrow" version of a band-stop filter. Lower curve: The shelving filter is characterized by the curve being flat in its active portion.

Crossover Frequency

In a crossover filter for a loudspeaker, this is the frequency or frequencies that determine what portion of the frequency spectrum will be distributed to the individual loudspeaker unit.

Attenuation

This is the magnitude by which a signal's strength is reduced, which is expressed in dB or the factor by which it is reduced.

Gain

This represents the magnitude by which a signal's strength is increased, which is expressed in dB or the factor by which it is increased.

Slope

Slope is the response curve outside the actual pass band, described as dB per octave.

Corner Frequency/Turnover Frequency

These terms indicate the frequency where two frequency ranges precisely overlap each other.

Q Factor

A filter's Q factor is an expression of the relative bandwidth.

$$Q = \frac{f_{res}}{b}$$

where
  f_{res} = resonance/center frequency
  b   = bandwidth

FILTER TYPES

The following gives a short description of the most commonly used filter types or equalizers.

Adaptive filter A filter that can adapt itself to a given spectrum, for example a noise spectrum, in order to be subsequently used to remove the noise concerned. Typically based on digital signal processing.

Bell The characteristics of the band-pass filter have the shape of a bell.

Band-pass filter Allows a specific frequency range to pass, but attenuates frequencies outside of it. This designation is however often used for filters which can both amplify and attenuate.

Band-stop filter Suppresses a specific frequency range, but allows frequencies outside of it to pass.

Constant Q/Constant bandwidth These have more or less become type designations for graphic equalizers, where the individual (band-pass) filters have the same bandwidth regardless of the magnitude of the amplification (or attenuation).

Graphic equalizer Consists of a number of band-pass filters, each of which can amplify or attenuate their part of the frequency range and which cover the entire frequency range collectively. The division into bands often follows the international standards for 1/1 or 1/3 octave, and possibly 1/2 octave. The "graphic" part of the term refers to the sliders or controls on the front panel of the device that provide an indication of the frequency response that is presumed to be attained.

High-pass filter Allows high frequencies to pass, but suppresses low frequencies. Also called LO-cut or bass-cut filters.

Low-pass filter Allows low frequencies to pass, but suppresses the higher frequencies. The term HI-cut filter is also used.

Notch filter Suppresses a very small range of frequencies. The bandwidth is a few percent.

Octave filter A band-pass filter with a passband of precisely an octave. An octave filter is typically designated by the center frequency for the range concerned. In sound technology, ten standard octaves are used as a rule, as set in the ISO standard (see *Standard frequencies*). The center frequency is the geometric average of the upper and lower cutoff frequencies:

$$f_c = \sqrt{f_u \times f_l}$$

The upper and lower boundaries are determined in the following manner:

$$f_u = 2 \times f_l \; ; \; f_l = \frac{f_c}{\sqrt{2}} \; ; \; f_u = f_c \times \sqrt{2}$$

1/3 octave filter The 1/3 octave filter, like the octave filter, has a constant relative bandwidth. There are also standardized center frequencies for these. The bandwidth is 22%.

Parametric equalizer A filter bank where the following parameters can be controlled: gain/attenuation, bandwidth, and turnover frequency.

Peak filter A filter that amplifies a very narrow band.

Presence A filter that amplifies in the frequency range around 2–5 kHz. This filter can add "nearness" or "presence" to acoustic images, particularly to voices.

Shelving filter This type of filter is often found in sound mixers as a high- or low-pass filter. It is characterized by the active area having a constant amplification or attenuation. In certain filters, the rise or drop can be maintained, whereas the transition frequency can be varied.

Sweeping filter See parametric equalizer.

Determination of Loudness

CHAPTER OUTLINE

Loudness is a measure of how loud an acoustic signal sounds and we will naturally be interested in measuring acoustic sound in a manner that expresses precisely the loudness experienced. Depending on the application there are several methods for doing this.

In principle, a weighted measurement like IEC A is an attempt to measure sound as it is subjectively perceived. This, however, is very approximate and not sufficiently precise as a measurement of loudness.

The basis of loudness measurement is, of course, the "equal loudness" curves. Each point (and thus frequency) on the same curve sounds equally loud, even though the sound pressure varies with the frequency. The curves express constant loudness regarding constant tones, and this loudness is expressed in phons. When we depart from pure tones and move towards complex and composite sounds, which consist of many frequencies of different duration, it is necessary to compute data in a more comprehensive manner.

ZWICKER'S METHOD

The most important method for the measurement and computation of loudness of noise was described by Eberhard Zwicker. The same method has also been standardized internationally.

The basis of this method is a frequency division into the so-called critical bands. As input data for the model, 1/3 octave spectra are used. However, frequencies under 280 Hz are added on an energy basis. In addition, it must

Audio Metering. DOI: 10.1016/B978-0-240-81467-4.10010-3

75

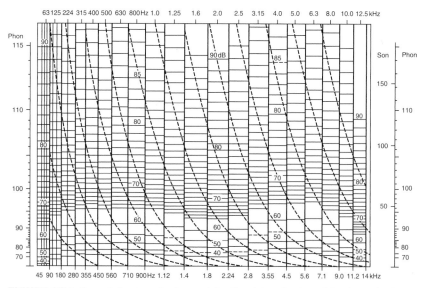

FIGURE 10.1 One of the ten charts used for the determination of loudness is shown here. This particular chart is called C8 and is used to determine loudness in a diffuse sound field for sound pressure in the 60–90 dB range.

be determined in advance whether the spectrum concerned is recorded in a direct, frontal sound field (the sound comes from only one direction, the front) or in a diffuse sound field (where all sound directions are equally possible). Loudness can then be computed manually by using one of ten charts as described below (see Figure 10.1). Alternatively, a number of instruments can measure loudness directly and provide the result in phons.

The principles are described as a standard in ISO 532-1975. This standard actually specifies two methods for determining a single numerical value of a given sound; the first method is based on physical measurements, such as spectrum analysis in octave bands, and the second method is based on spectrum analysis in 1/3 octave bands.

Zwicker's articles that formed the basis for the standard can be found in Acoustica, No. 10, p. 303 (1960) and in the Journal of the Acoustical Society of America, No. 33, p. 248 (1961).

Procedure

Manual determination of loudness by Zwicker's method is performed in three steps.

First, the proper chart is selected. Are we measuring a direct frontal sound field or a diffuse sound field? Then the chart is selected where the highest value that occurs in the 1/3 octave spectrum can be plotted on the chart. The 1/3 octave values from 45 to 90 Hz are added together to one value. The same applies for the 90 − 180 Hz values and the 180 − 280 Hz values. These values are then plotted on the measurement chart.

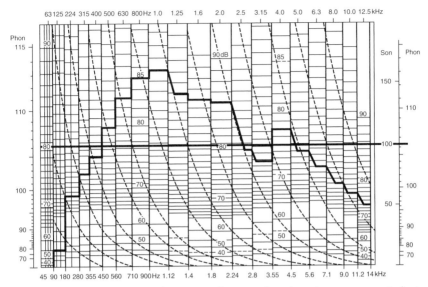

FIGURE 10.2 The measured 1/3 octave values are plotted on the measurement chart. Lines are drawn between the values from left to right. For a rising level, vertical lines are used. For a falling level, follow the dotted lines. Finally, a horizontal line is drawn where the area that is inside the curve above the line is of the same magnitude as the area that is outside the curve under the line. The line is extended to the scale on the right side. The point where it intersects the scale indicates the result.

The next step consists of connecting the levels plotted so that a following higher frequency band is connected with a vertical connecting line if the level in this next band is higher than the value in the previous one. If the next band has a lower value, the line follows a parallel to the slanted lines. (See Figure 10.2).

The third step consists of drawing a transverse line such that there is just as much curve area above the line as under it. The line is extended out to the scale on the sides of the measurement charts. The point of intersection then indicates the loudness value.

As mentioned earlier, the method has been standardized under the auspices of the ISO. The other charts can be found in the standard.

Corrections to Zwicker's Method

Some researchers have made a revision to Zwicker's method. This is not currently a standard, however. A description can be found in the following article:

Moore, B.C.J., Peters, R.W., and Glasberg, B.R. A revision of Zwicker's loudness model. Acta Acoustica, vol. 82, pp. 335–345 (1996).

In practice, there are only insignificant differences between the original and revised versions of Zwicker's method.

STEVENS' METHOD

A somewhat simpler method was designed by S.S. Stevens. It is based on 1/1 octave analysis. It is not as precise as Zwicker's method, but it is also included in ISO 532-1975.

Each octave band is converted to a loudness index based on a measuring chart similar to the example shown in Figure 10.3. It requires differentiating between a direct, frontal sound field and a diffuse sound field. There are only two charts to choose from.

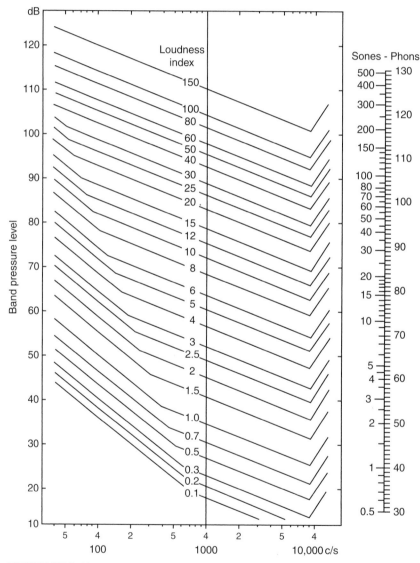

FIGURE 10.3 Measurement chart used in the manual calculation of loudness as per Stevens' model. This is the calculation chart for a diffuse sound field.

The different loudness indices are added using the following expression:

$$S_t = S_m + 0.3(\Sigma S - S_m)$$

where
S_m = the highest loudness index occurring in the measurement concerned
ΣS = the sum of the loudness indices in all bands
S_t = the total loudness expressed in sones

DOLBY L_{eq}(M) MOVIE LOUDNESS/ANNOYANCE

Dolby's method, which was mentioned earlier when discussing frequency weighting, claims to measure loudness. However, the value obtained is rather "annoyance." It is a method that was developed to measure the sound level of movies, particularly trailers and advertisements.

The measurement of a film's sound is made electrically since it is presumed that a film will be played in a cinema with a calibrated sound system. The sound from each track is frequency-weighted and filtered with CCIR 468 (level-corrected to 0 dB at 2 kHz). All signals are detected and then summed on an energy basis to a single value. An average is taken over the total length of the film. The final value arrived at in the measurement is designated L_{eq}(m). This involves a dose measurement that adds the levels on an energy basis, since the measurement must relate to the sound field in the interior of the cinema.

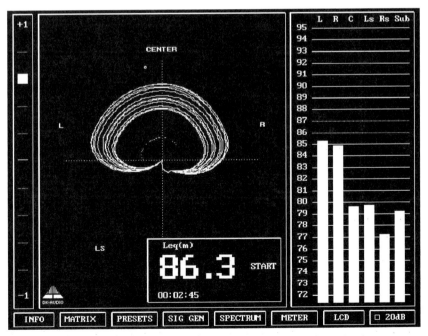

FIGURE 10.4 The L_{eq}(m) standard has been implemented on, among other instruments, the DK-Audio MSD600.

When a test film from Dolby called Cat. No. 69 is played in a cinema, this meter would show that $L_{eq}(m) = 82$ dB.

ITU-R BS.1770 — LOUDNESS IN BROADCAST PROGRAMS

In broadcasting it is a still increasing problem that viewers watching television programs often complain about audio loudness jumps at every commercial break. Television commercials have unfortunately been infamous for their high compression and loud play-out. Mixed with other program material the level has been evaluated as very uneven. The problem could not be solved with existing standard metering and the traditional level setting alone.

A lot of effort to address this issue was made in standards organizations as well as in private companies around the years 2000–2005. Most of these interested parties were gathered in an ITU-initiated workgroup created to address this problem. Large scale tests were initiated to find a practical solution for measurement and control of loudness. Many old and new algorithms and meters were tested against real program material. Fortunately, the outcome of the ITU work was presented in 2006 as a surprisingly simple solution that has become the ITU-R BS.1770 recommendation.

The basis for this recommendation is a weighting filter — the RLB curve (see Chapter 9 on frequency weighting) and a L_{eq} measurement. The RLB (revised low-frequency B-curve) proved to be the best solution for continuous mono signals. As broadcast programs may contain not only mono, but also two-channel mono, stereo, and 5.1 surround sound, a further study was made to provide one single loudness value regardless of the number of channels.

The presence of a head in the sound field has an influence on the spectrum. The acoustic effect of the head modeled as a rigid sphere was taken into account by adding pre-filtering to compensate by raising the levels above 2 kHz by 4 dB.

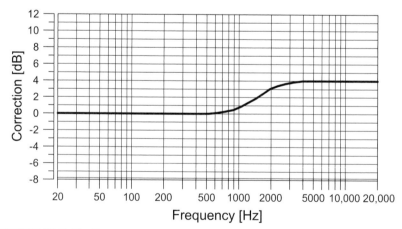

FIGURE 10.5 The ITU 1770 pre-filter added to all channels prior to summation in order to compensate for the acoustic effect of the head.

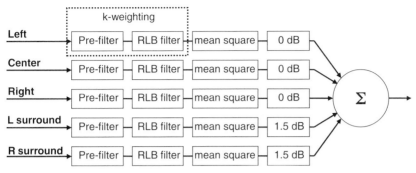

FIGURE 10.6 Block diagram of multichannel loudness algorithm (ITU 1770).

Humans' awareness of sounds coming from the rear has also been taken into account by adding approximately 1.5 dB (a factor of 1.41) to the left surround and right surround, respectively, prior to the summation. The complete processing can be seen in Figure 10.6.

The additional weighting together with the original RLB curve received a new name: k-weighting. In this respect, loudness was indicated on a scale with reference to a full scale maximum. The loudness level thus became LKFS, meaning level, k-weighted, with reference to full scale. A unit of LKFS is equivalent to a dB.

The recommendation "ITU-R BS.1770 — Algorithms to measure audio programme loudness and true-peak audio level" has been a major breakthrough on loudness measures in the field of broadcast. It provides the possibility of defining a target level for the instantaneous loudness as well as of the full length of the program based on an adequate loudness measure. (It is however noted in the recommendation that in general it is not suitable for estimating the loudness level of pure tones.)

EBU R-128 — GATING

The European Broadcast Union accepted the ITU 1770 loudness algorithm. However, in the recommendation EBU R-128, the program loudness estimation has been taken a bit further:

The ITU 1770 algorithm measures rather long term or continuous loudness and does not take into account certain types of programs that may contain larger sequences with very low-level audio. This could be a TV program about wildlife having a narrator with long views overlooking the savannah while the wind is silently whispering in the grass in between the narration. If the level of the total program is averaged to reach the target level then the narrating part will be gained too much and become too loud on air. Additionally, commercials could be produced "creatively" including silence in order to become louder in the non-silent parts.

The EBU workgroup P/LOUD adopted the ITU algorithm for loudness estimation. One thing that was changed is that the named LKFS scaling was

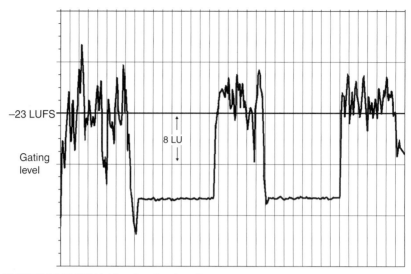

FIGURE 10.7 EBU-initiated gating function.

substituted by Loudness Unit relative to Full Scale (LUFS). However, in addition to this a gating function was introduced. This gating function halts the calculation whenever the loudness level is 8 LU below the target program loudness level. As mentioned, the basic loudness algorithm is the same as in the ITU recommendation but the gating function prevents sequences in the program from exceeding the target level. The practical loudness reading includes long-term, short-term and instantaneous loudness readings (more about this in Chapter 14 on Loudness Metering).

Bibliography

Zwicker, E. Ein Verfahren zur Berechnung der Lautstärke (A procedure for calculating loudness), Acustica, vol.10, pp. 304—308.

ANSI S3.4-1980 (R1997). Procedures for the computation of loudness of noise.

Moore, B.C.J., Peters, R.W., and Glasberg, B.R. A revision of Zwicker's loudness model. Acta Acoustica, vol. 82, pp. 335—345.

ISO 532:1975. Acoustics — Method for calculating loudness level.

ITU-R BS.1770 (2006). Algorithms to measure audio programme loudness and true-peak audio level.

EBU Recommendation R 128 (2010). Loudness normalisation and permitted maximum level of audio levels.

Skovenborg, E., Nielsen, S. Evaluation of different loudness models with music and speech material. AES 117th Conv. 2004. Preprint 6234.

Soulodre, G. A., Lavoie, M. C., and Norcross, S. G. The Subjective Loudness of Typical Program Material, AES 115th Conv. 2003. Preprint 5892.

Soulodre, G. A. Evaluation of Objective Loudness Meters, AES 116th Conv. 2004. Preprint 6161.

Soulodre, G. A., and Lavoie, M. C. Stereo and Multichannel Loudness Perception and Metering, AES 119th Conv. 2005. Preprint 6618.

Chapter eleven

Characteristics of Level Meters

CHAPTER OUTLINE

The characteristics of individual instruments are normally defined and set in international standards. Some of the most important standards are IEC 60268-10, IEC 60268-17 and IEC 60268-18. In addition, recommendations may be provided by the ITU or the EBU. Alternatively, there can be national standards which apply, such as ANSI or "Technische Pflichtenhefte der öffentlich-rechtlichen Rundfunkanstalten in der Bundesrepublik Deutschland" (Publication of technical requirements for public broadcasters in Germany).

DEFINITIONS AND REQUIREMENTS

In this chapter a number of general definitions and specifications will be reviewed together with examples of related specifications. The specifications here all refer to the IEC standards.

Reference Indication

Analog instruments The instrument must have a marking on its scale that shows the maximum level allowed in the system to which the instrument has

Audio Metering. DOI: 10.1016/B978-0-240-81467-4.10011-5

been connected. This marking can be expressed in percent (for example 100%), dB (for example 0 dB), or Volume Units (for example 0 VU). This does not necessarily involve an absolute level.

Digital instruments The reference indication corresponds to full scale. The indication is normally marked 0 in terms of dB.

Reference Input Voltage

Analog instruments This is the reference signal necessary to reach reference indication and is described by the RMS value of a 1000 Hz sinusoidal signal.

IEC Type I: If nothing else is specified, the voltage is 1.55 V. It can also be expressed in dB relative to 0.775 V. For example, +6 dB re 0.775 V or +6 dBu.

Volume indicators, VU (as per IEC 60268-17): The applied voltage for reaching the reference position is 1.288 V.

Division of The Scale

Analog instruments This is the graphical display of the signal concerned.

IEC Type I: −40 dB − +3 dB (minimum).

IEC volume indicator: −20 − +3 VU

Digital instruments The graphical display can either be an incremental dot or bar graph. Numerical values may indicate headroom or a reference if the input is analog.

Scale markings:

0 dB to −20 dB	numbers per 5 dB and minor tics every 1 dB
−20 dB to −40 dB	numbers per 10 dB and no minor tics
−40 dB to −60 dB	numbers per 10 dB and a minor tic at −45 dB

Amplitude Frequency Response

This is the difference, expressed in dB, which occurs between the levels indicated for a given frequency and the indication that occurs at a reference frequency, normally 1000 Hz.

Analog instruments IEC PPM: 31.5 Hz − 16 kHz ± 1 dB

Digital instruments IEC digital: 20 Hz − 20 kHz ± 0.5 dB

Dynamic Response

This expresses what the instrument displays for a 5 kHz tone burst of different durations (100, 10, 5, 1.5 ms) in relation to a continuous 5 kHz sinusoidal tone (or >10 kHz for very short responses like 0.5 ms). The requirements are expressed by a curve or in a table.

Delay Time

Analog instruments The time it takes from when a reference voltage is applied until the display reaches 1 dB under the reference value.

IEC PPM: Less than 300 ms

Digital instruments Delay time is the time interval between the application of the reference input signal and the moment when the indicator passes a point 1 dB below reference indication. This delay is normally related to processing time.

IEC digital: Less than 150 ms

Integration Time

This is the duration of a tone burst at a reference level that brings the indication to a given point under the reference value.

Analog volume indicator 300 ms to reach 99% of the reference value
Analog and digital PPM instruments 5 ms results in an indication 2 dB below the reference value

Overswing

This represents the maximum indication above a reference value that occurs when a 1000 Hz signal is applied to the instrument at the reference level.

Analog and digital instruments ≤ 1 dB

Return Time

Analog and digital instruments The time it takes for the indication to fall to a defined point below the reference value when an applied constant signal at the reference level is removed.

IEC PPM and digital: 1.7 s $\pm$ 0.3 s / 20 dB

Reversibility Error

Analog instruments The difference in the indication when an applied asymmetric signal is phase-inverted.

IEC Type I: ≤ 1 dB

Input Impedance

This is the internal impedance of the instrument in its entire active frequency range.

Analog, PPM: >10 kΩ
Analog, volume indicator: 7.5 kΩ

Distortion Introduced by the Peak Program Level Meter

This represents the total harmonic distortion that is caused by the presence of the instrument.

IEC Type I: $<0.1\%$

Overload Characteristic

The expresses the maximum input signal (sinusoid) that the instrument can handle without it subsequently altering the specifications of the instrument.

IEC Type I: >20 dB for a 5 second signal
>10 dB for a continuous signal

Supply Voltage Range

Analog and digital instruments The maximum allowable variation in supply voltage for a given deviation in the indication.

The chapters which follow will describe the relevant instruments in more detail.

Chapter | twelve

The Standard Volume Indicator (VU Meter)

CHAPTER OUTLINE

The Standard Volume Indicator (SVI) (also commonly known as a VU meter) was originally developed by the Bell System around 1940. It was standardized in 1942 (American Standards Association (ASA), C16.5-1942: American Recommended Practice for Volume Measurements of Electrical Speech and Program Waves), and has ever since been the most used — and perhaps most misused and misunderstood — analog level meter in audio. The latest reaffirmation of the standard was issued by the IEC in 1990. The ANSI standard was withdrawn in 1999.

BASIC SPECIFICATIONS

The original Standard Volume Indicator consisted of a full wave rectifier and a galvanometer (i.e., an electromechanical transducer). Hence the meter is based on an average reading of the program signal, not peaks. According to the standard the scale ranges from -20 VU to $+3$ VU. The integration time, that is, the time it takes the deflection to reach from the bottom to the reference point of the scale (0 VU), corresponds to 300 ms $\pm$ 10%. The overshoot must be within $1-1.5\%$.

"VU" was originally denoted as a term designated to indicate volume. It was not meant to be a unit. However, later standards and practice have translated the VU as "Volume Unit."

Audio Metering. DOI: 10.1016/B978-0-240-81467-4.10012-7

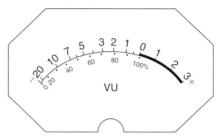

FIGURE 12.1 The VU-scale on the Standard Volume Indicator ranges from −20 to +3.

ATTENUATOR

An important part of the Standard Volume Indicator is the attenuator circuitry connected with the meter. This passive electronic network determines the impedance and sensitivity of the complete meter. The reading of the program volume actually takes place on the attenuator − not the meter!

As can be seen from Table 12.1, the standard attenuator operates in steps of 1 dB. At 0 dB attenuation the meter will reach the reference point on the scale (0 VU) when connected to a source level of +4 dBm. As the appendage "m" indicates, the reference is 1 mW. This corresponds to a voltage of 0.775 V across a load of 600 Ω; hence the +4 dBm corresponds to 1.288 V.

IMPEDANCE

The input impedance of the attenuator (including the galvanometer) is 7500 Ω. A load resistor of 600 Ω can be put across the input terminals. So when terminated with 600 Ω (impedance matching) and supplied by a constant signal of +4 dBm, the reading of the meter is 0 VU. The initial standard was not restricted to 600 Ω systems. If used with other impedances the power level would however change and the calibration of the meter would be different from that of a 600 Ω system.

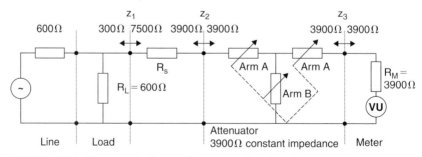

FIGURE 12.2 VU-meter circuit providing attenuation for indication of reference levels above +4 dBm.

TABLE 12.1 Values for the Standard Volume Indicator attenuator.

Attenuation dB	Level VU	Arm A Ohm	Arm B Ohm
0	+4	0	Open
1	+5	224.3	33801
2	+6	447.1	16788
3	+7	666.9	11070
4	+8	882.5	8177
5	+9	1093	6415
6	+10	1296	5221
7	+11	1492	4352
8	+12	1679	3690
9	+13	1857	3166
10	+14	2026	2741
11	+15	2185	2388
12	+16	2334	2091
13	+17	2473	1838
14	+18	2603	1621
15	+19	2722	1432
16	+20	2833	1268
17	+21	2935	1124
18	+22	3028	997.8
19	+23	3113	886.3
20	+24	3191	787.8

THE CONFUSION

The problem of understanding the reading of the SVI or VU meter is that the scale does not indicate VU.

Notice: **When**
- the line is terminated with 600 Ω
- and the reading on the attenuator is "+4 VU" (the attenuation is 0 dB)
- and the reading of the meter is "0"

then this together indicates that the **volume level** of the program is "4".

If the program level is raised, then more attenuation is needed to align to the 0 VU marking on the meter; thus the volume of the signal is higher.

The confusion is that a 0 VU reading on the scale of a meter meeting the standard in a 600 Ω system can indicate any volume level between +4 and +24 depending on the setting of the attenuator (if this attenuator exhibits a 0 to 20 dB attenuation).

THE SVI AND PEAKS

Because the SVI — or VU meter — is an average-reading instrument it reacts relatively slowly and peaks in music signals will often not be registered at all. In practice, the peaks can be 6—12 dB above the actual reading of the instrument. If program material with impulsive content is modulated to a display around 0 VU, overloading can quite easily occur on tapes, amplifiers, transmitters, etc.

To prevent this, a genuine VU meter is provided with a so-called lead. This is an amplification circuit that can provide an increased reading on the meter. If this lead is not present, then correctly modulated impulsive type music would only show deflection at the bottom of the scale. Unfortunately, for years virtually no mixing desks equipped with VU instruments have included this lead.

THE "MODERN" VU METER

The understanding and use of the (still remaining) VU meters today includes only a few of the initial intentions. Basically, the scale (including the background color: US Postcard Yellow) and the integration time of 300 ms are intact. The reference today is no longer power but voltage. So the 0 VU deflection is reached for the voltage of 1.288 V. In most applications the attenuator is totally forgotten and the scale is translated to indicate the VU as a unit. Usually the input impedance is high (i.e., ≥ 7.5 kΩ). While not included in any standard, a very practical LED overload indicator can be included with the VU meter.

CALIBRATION USING THE VU METER

The VU meter has been widely used in the calibration of, for instance, tape-based media. The precise calibration values (magnetization levels) have changed over the years with the change in tape sensitivity. This is why the VU reading in these cases is related to the media rather than to volume or voltage. The individual references have to be looked up. See Chapter 21, "Standards and Practices."

Bibliography

American Standards Association (ASA), C16.5-1942: American Recommended Practice for Volume Measurements of Electrical Speech and Program Waves.

Hertz, B. New CCITT "Three-level test signal": The latest aspect in the VU—PPM debate. AES 80th Conv. March 1986. Preprint 2357.

IEC 60268-17 ed1.0 (1990) Sound system equipment. Part 17: Standard volume indicators.

McKnight, J.G. (Jay). Some Questions and Answers on the Standard Volume Indicator ("VU meter") (revised 2006) http://www.aes.org/aeshc/pdf/mcknight-qa-on-the-svi-6.pdf.

Pavel, E.A., Gastell, A., Bidlingmaier, M. Über Vergleichende Messungen mit dem Volumenmesser und dem Spitzenwertmesser bei der Kontrolle von Rundfunkübertagungen. FTZ, Heft 4. 1955.

Wilms, H. A. O. VU- versus PPM-indicators. The end of a continuous misunderstanding. AES 56th Conv. March 1977. Preprint 1221.

Chapter | thirteen

Peak Program Meter — PPM

CHAPTER OUTLINE

THE PPM

The PPM (Peak Program Meter) is more level orientated and thus works significantly faster than the VU meter. The instrument's time constant, the integration time, that is, the time it takes the reading to reach a point 2 dB below the reference indication, is 5 ms. In earlier standards the integration time was listed as 10 ms. However, this is rather a question of definition. It still takes 10 ms to reach a reading 1 dB below the reference indication. In any event, the short integration time ensures that peaks of short duration basically can be read at (nearly) the correct level.

In order for the eye to be able to register the indication, the fallback time (or return time) is made relatively long: 1.5 sec. per 20 dB. The PPM meter is so fast (30 times faster than the VU meter) that in practice only peaks that are 3 dB higher than the displayed value will exist. Thus, the analog PPM nowadays is often called QPPM (Quasi Peak Program Meter) due to the fact that it is still too slow to show the true peaks of the signal.

In a digital PPM, problems can arise at specific signal frequencies that are a subdivision of the sampling frequency even if the value displayed is computed on the basis of all samples. Special attention is not paid to this in the standard since the error can arise for only a very few frequencies.

Audio Metering. DOI: 10.1016/B978-0-240-81467-4.10013-9

This error is also normally smoothed out since averaging is performed over several samples.

However, in newer broadcast formats (like EBU R-128) peaks are allowed to reach a level 1 dB below full scale. In order to avoid any errors due to the occurrence of very short peaks and an unfortunate relationship between audio frequency and sampling frequency, it is recommended to measure "True Peak" by the use of (at least) four times oversampling.

DIN SCALE

The DIN scale is characterized by the "0 dB" marking situated at the edge of the "red area," which is the normal designation for the overload region. This modulation corresponds to a voltage of 1.55 V (+6 dBu). The scale ranges from −50 to +5 dB. A marking for the test level lies at −9 dB on the scale (see the "Test Level" section below). In addition to dB markings, the scale can be furnished with a percentage scale, where 0 dB = 100% (level), −6 dB = 50% and −20 dB = 10%.

The DIN scale has the advantage that the 0 dB marking is positioned at the reference indication (i.e., at the edge of the "red" range). If at some point in time a different voltage reference is chosen, as in a digital recording, the scale can still be used.

The DIN scale was originally standardized in DIN 45406. However, this has subsequently been replaced by a type I meter in the IEC 60268-10 standard.

NORDIC SCALE

The Nordic broadcasters agreed that it would be practical to use a scale that was calibrated in dBu. This scale is called the "Nordic scale."

The edge of the red area lies at +6 dBu (with 0 dBu = 0.775 V). At a minimum, the scale covers a range from −36 to +9 dBu; however, certain implementations exist in which it runs from −42 to +12 dBu. The test mark lies at 0 dBu, 6 dB below the edge of the red area. Note that full modulation to the edge of the red area is attained at a voltage of 1.55 V for both the DIN scale and the Nordic scale.

BBC SCALE

The BBC developed a scale that runs from 0 to 7. The scale was designed in this manner for reasons of clarity. The number "4" is located at the middle of the scale. At this level, the instrument's indicator is in a vertical position, and the voltage is 0.775 V (0 dBu).

The standard requires there being approximately 4 dB between each of the digits 1−7. The number "5" corresponds to 0 on the VU-meter scale (for a constant tone and no attenuation added). Meters exist with the BBC scale that (outside the standard) use approximately 6 dB in the 1−2 and 2−3 intervals.

TEST LEVEL

In a broadcasting context, the concept of a "test" or "test level" is used. What this refers to is one or more tones with a well-defined frequency, level, and duration.

In connection with the exchange of video or audio programs, a test tone of 1 kHz is normally used in the Nordic countries, which is stated as being recorded as 6 dB below full modulation or, rather, 6 dB below the reference indication.

What can seem a little strange is that two of the PPM instruments mentioned above provide a "Test" marking, but at two different positions on the scale in relation to the red area! The explanation is quite amusing in historical terms; however, the result can be a bit unmanageable in practice. The test tone concept originates from the era when AM radio transmitters came into use. A test signal was desired that contained the same energy as "Gewöhnlisch Tanz-musik," that is, normal dance music, which on the average corresponded to a 35% modulation of the transmitters. This 35% corresponds to 9 dB under full signal (9 dB under 100% modulation). This is the reason for the DIN scale's test point at −9 dB.

When the Nordic scale was conceived it was taken into consideration that the PPM meter, with its integration time of 10 ms, in some situations could indicate a value that was 3 dB too low. In other words, level peaks up to 3 dB above +6 dBu can slip through even if the display never moves into the red area. As the test point lies 9 dB below the maximum, the result becomes 0 dBu! In the Nordic countries, the test level is thus 6 dB below the indicated full modulation.

When providing audio or video programs with test tones, as much information as possible should always be given so there will be no misunderstandings over the test level. If the material is tape based the corresponding magnetic flux should be noted. In the digital domain the levels should be noted with reference to full scale.

On a BetaCam, it has become normal to use devices with a VU scale. The test level is placed at 0 VU. Despite the fact that the scale only goes to +3 VU, it is customary to make room for the obligatory 9 dB above test. The VU meter is then used only for adjusting the test tone.

EXTRA FUNCTIONALITY

Instrumentation is often furnished with practical additional functionality:

Integration Time: Fast

The standardized integration time of PPM instruments of 5 ms (to reach a level 2 dB below that of a constant tone) can be supplemented with a "Fast" option, which is normally 0.1 ms. This capability can be useful with both analog and digital recordings.

Peak Hold

This function can show what the maximum level has been during a given time period. It is typically released by a reset button or, alternatively, it can have a hold time of some seconds.

Additional Gain

Here, the signal is amplified before the reading is displayed; this is typically 20 dB for PPM instruments. This function can be applied when working with a large dynamic range or when program material with varying modulation is to be assessed.

Peak Indication

Certain instruments can be equipped with an additional LED indicating whether peak values exceed the maximum permitted level or the maximum of the scale. This function can be beneficial if the time constant and level is known, which is seldom the case in non-professional meters. However, if a time constant of 5 ms (as a PPM instrument) or shorter is provided, then this can be a quite practical feature.

Bibliography

IEC 60268 Sound System equipment. Part10: Peak programme level meters.
IEC 60268 Sound System equipment. Part18: Peak programme level meters — Digital audio peak level meter.
EBU Recommendation R 128 (2010). Loudness normalisation and permitted maximum level of audio levels.

Loudness Metering

CHAPTER OUTLINE

Loudness is defined as the subjectively assessed level of sound. The task of an instrument named a loudness meter is more or less to emulate human hearing. The problem, however, is that the hearing and the assessment of sound to some degree is individual from person to person. Hence a true loudness meter is very difficult to define. Level, dynamic range, frequency range, the direction of received sound, and the character of the sound itself will have an effect on the perceived loudness. Further it is not possible to express the loudness of an electrical signal; it is the acoustic sound received and perceived by the listener that should be the basis for the measurement or calculation of loudness. So in reality the truth is impossible to reach. It is rather a question of defining a system that exhibits minimum error, a system that is satisfactory to most listeners.

For years there have been systems striving to measure loudness, the Zwicker method being the most significant. This method is still valid regarding humans' assessment of noise. Recently, intensive research has taken place trying to define the optimum system for audio production and broadcast. This has led to new recommendations (ITU-R BS.1770, EBU R 128) that may develop further in the future. This chapter tries to provide an overview of the most important existing and new possibilities for loudness metering. Most attention is paid to the EBU meter, as this must be regarded as the state-of-the-art.

Audio Metering. DOI: 10.1016/B978-0-240-81467-4.10014-0

There are a number of loudness measuring systems that have been in use for audio production and broadcast for a long time. Since the Standard Volume Indicator or VU meter was developed and standardized, alternative ideas have found their way into audio production facilities.

DORROUGH LOUDNESS METER

The loudness meters from Dorrough Electronics have been widely used both in music production and broadcast since their introduction in the early 1970s. From that time these meters have been manufactured as stand-alone units for analog and digital audio or as plug-ins for digital audio workstations (DAWs).

The meters essentially simultaneously display the instantaneous peak level of the signal as well as the average program level on the LED scale, which for basic use has a linear range of 40 dB. The difference between the readings expresses the "density" of the program. Each channel is measured individually.

The peak acquisition period is 10 μs to full scale (measured with a 25 kHz sine input). The peak decay period is 180 ms, from full scale to all LEDs off. The loudness impression is derived from an average reading of the program signal. The time constant for the average reading is 600 ms.

Most systems in the line facilitate two-channel/stereo signals. Inter channel phase relations can be monitored. The reading can be changed from LR to MS (Sum/Diff modes). Special automatic zooming of scales for calibration is provided (Dorrough Window Expansion Mode). Various alarms can also be set (e.g., for over/under level and for phase errors).

DOLBY METERS

Dolby Labs is a major provider of audio equipment and systems primarily to the film industry and broadcast. Over the years several Dolby inventions have become de facto standards. Dolby metering systems have been developed to meet the demands from the industry.

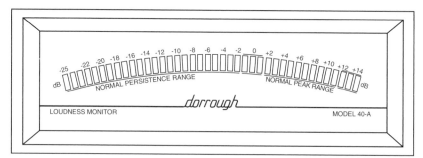

FIGURE 14.1 Dorrough Loudness Meters provide simultaneously a reading of the peak and the average program level.

Model 737 Soundtrack Loudness Meter — L$_{eq}$(m)

This meter has been developed for the measurement of L$_{eq}$(m), primarily on film (see Chapter 10 on Determination of Loudness). While this unit has actually been withdrawn, a large number of Model 737 meters are still in service, as L$_{eq}$(m) has become a standard for the measurement of commercials and trailers in many countries. It should be mentioned that the measure is more annoyance than loudness.

LM100 Broadcast Loudness Meter

The Dolby LM100 Broadcast Loudness Meter was developed as a tool for measuring the subjective loudness of dialog within broadcast programming. The background for this is the fact that dialog can be regarded as the "anchor element" of a program; lining up dialog levels provides a better equality of loudness within and between programs.

The LM100 employs the proprietary technology Dialogue Intelligence™ to measure the perceived loudness of dialog in a complex program. The early versions utilized L$_{eq}$(A) for this measurement. However, the newer versions also have implemented the ITU-R BS.1770 algorithm as well (discussed further later in the chapter). Further, the instrument can determine the un-weighted peak and a range of other information about the signal. The unit can simultaneously display the incoming dialog normalization (dialnorm, see Chapter 17 on Dynamic Scales) value of a Dolby Digital program or any program within a Dolby E bitstream for direct comparison with the actual measured value.

A set of user-definable alarms and monitoring functions can inform an operator of input loss, signal clipping, over modulation (LM100-NTSC version), high or low signal levels, silence, and incorrectly set dialnorm values.

Dolby Media Meter 2

The Dolby Media Meter is a software tool that measures loudness in programming for broadcast, packaged media, cinema trailers, VOD, and games. Beside the features found in the LM100 the Media Meter can perform the measurement of L$_{eq}$(m). This software meter should be regarded as a substitute for the Model 737. The Dolby Media Meter 2 runs on all major platforms.

FIGURE 14.2 An example of the display read out on the Dolby LM100.

ITU-R BS.1771 (2006)

With the digital age entering broadcast the International Telecommunications Union realized that a number of problems in transmission needed solutions. A larger dynamic range was now available. The transmission would cover formats from mono to surround. Downstream conversion to lower bit rates would create alteration of peak levels. However, the most serious problem was the different levels perceived by the listener when zapping between the channels or when the same channel was changing between different content: the loudness problem.

A workgroup facilitated the psycho-acoustic testing of different already available and new loudness meters and algorithms. The "best fit" for real audio mono samples proved to be a L_{eq} of the RLB-weighted signal. After reaching this conclusion, the work was expanded to include the evaluation of program contents in stereo and surround as well. This involved a pre-filter to compensate for the presence of the head and a gain to surround channels due to the special awareness effect of sound coming from the rear. The total weighting of the signal (frequency and level) got the name k-weighting ("k" being an available symbol for this purpose). The algorithms are defined in the recommendation ITU-R BS.1770 from 2006.

The ITU Loudness Meter

The advantage of the "winning" algorithm is that it is not proprietary to any private company and the meters can be manufactured by everybody. The meter specifications are explained in ITU-R BS.1771: Requirements for loudness and true-peak indication meters (2006).

The introduction of the Loudness Unit, LU, was a consequence of the practical implementation of this meter. In this standard the LU is a relative unit. The absolute loudness level at the reference indication can be defined elsewhere. The absolute level of the program loudness is defined by LKFS, meaning the level of the k-weighted signal with reference to full scale. Thus, the reference indication or the target loudness (0 LU) at an absolute level must be stated in LKFS.

Additionally, the absolute peak level (true peak) had to be considered. In digital systems oversampling must be used. At four times oversampling the worst case under-read is in the range of 0.6 dB. At eight times oversampling the worst case under-read is in the range of 0.15 dB. For the purpose of the standard, four times oversampling was chosen for the true peak reading.

DEFINITIONS

Loudness Unit (LU) The loudness unit is the scale unit of the loudness meter. The value of the program in loudness units represents the loss or gain (dB) that is required to bring the program to 0 LU; e.g., a program that reads −10 LU will require 10 dB of gain to bring that program up to a reading of 0 LU.

Type I electronic display Electronic display with resolution of one or more segments per loudness unit.

Type II electronic display Electronic display with resolution of one segment per three loudness units.

The following features are required to fulfill the recommendation:

General requirements	The loudness display reading must not vary by more than 0.5 Loudness Units when the signal polarity is reversed.
Common requirements for program loudness displays	The loudness display shall be calibrated in loudness units. Loudness of a stereo or multi-channel sound program shall be shown by a single display. (This does not prevent meters from also displaying individual channel loudness.)
Requirements for program loudness display — mechanical type	A mechanical loudness meter display shall have a nonlinearity of not more than 1% of full-scale deflection over its operating range.
Display requirements — Optional peak level indicator on loudness meter	The threshold for overload indication shall be −2 dB re full scale digital input. The overload indicator shall activate if the true-peak digital audio level exceeds the threshold. Once the indicator light is activated it shall remain activated for at least 150 ms after the signal has fallen below the threshold.

In addition to these mandatory requirements a number of optional facilities are proposed. One of these is that the loudness meter may have at least two operating modes: F mode (fast) and I mode (integrating). Further, the integrating mode may have a start/stop button or switch.

TC ELECTRONIC LM5/LM5D

TC Electronic has played an active part in the work for establishing standards for loudness measures. The ideas and research behind the LM5 software-based meter has impacted the EBU recommendation and it is expected that EBU terminology will be reflected in future updates of the meter. The initial versions of the meter operate with LU (Loudness Units as per ITU-R BS.1770), and LFS (loudness with reference to full scale, comparable to LKFS or LUFS).

The LM5 has the basic circular-shaped graphic. Some of the descriptors displayed by the meter are **Short-Term Loudness** (outer ring)

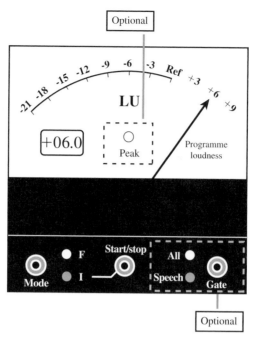

FIGURE 14.3 Example of program loudness display, mechanical type given by ITU-R BS.1771.

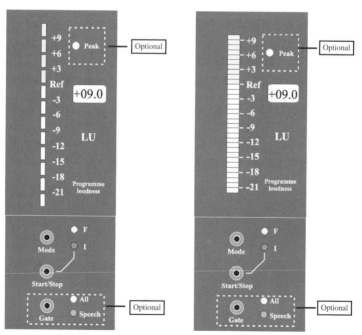

FIGURE 14.4 Example of program loudness level display, optoelectronic Type I (left) and Type II (right) given by ITU-R BS.1771.

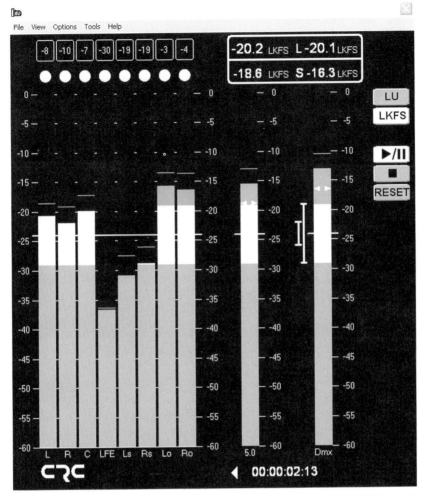

FIGURE 14.5 This display, of the ITU-R loudness meter developed at CRC, is the result of CRC's collaboration with CBC/Radio-Canada. The scale can display either LKFS or LU.

and **Loudness History** (radar). Additionally, the true peak levels are displayed.

The LM5D displays long-term statistical descriptors that describe an entire program, film, or music track. **Center of Gravity (CoG)** indicates the average loudness of a program, and is directly operational. If, for instance, a broadcast station is operated at an average loudness level of −23 LFS, and a commercial has its Center of Gravity measured at −19.5 LFS, the program should be attenuated by 3.5 dB before transmission for a best fit.

Consistency indicates the loudness variations inside a program. Center of Gravity ranges from −80 LFS to +12 LFS, while Consistency ranges from −40 to 0 LU.

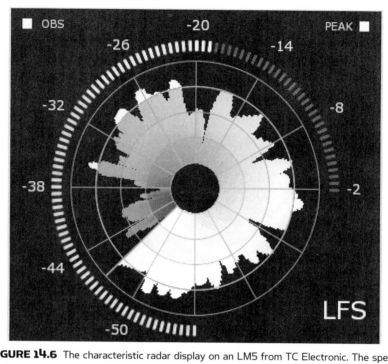

FIGURE 14.6 The characteristic radar display on an LM5 from TC Electronic. The speed of the radar can be set.

EBU R 128 LOUDNESS METERING

The European Broadcast Union wanted to implement the loudness measure already accepted by the ITU. In addition to this a gating technique was considered as well as a new descriptor called "Loudness Range." The new practices should replace the existing standard for PPM and at the same time exploit the dynamic range provided by a digital system.

EBU R 128 was published in autumn 2010. This recommendation introduced three descriptors: Program Loudness, Loudness Range, and Maximum True Peak Level. It also defined target levels as well as specifications for a meter to display the measures.

Program Loudness

Program Loudness is determined using k-weighting (as per ITU-R BS.1770) and averaging over the total length of the program. However, this includes a gating 8 LU below the target level. Whenever the program is below this gating level the loudness calculation is paused. (See Chapter 10 on Determination of Loudness.)

Loudness Range (LRA)

Loudness Range (LRA) is defined in EBU Technical Document 3342. Originally this descriptor was developed by TC Electronic (and named

"Consistency"). LRA is defined as the difference between the estimates of the 10th and the 95th percentiles of the distribution. The lower percentile of 10%, can, for example, prevent the fade-out of a music track from dominating the Loudness Range. The upper percentile of 95% ensures that a single unusually loud sound, such as a gunshot in a movie, cannot by itself be responsible for a large Loudness Range.

The computation of the Loudness Range is based on the statistical distribution of measured loudness using a sliding analysis window with a length of 3 seconds for integration. An overlap between consecutive analysis windows is used to retain precision of the measurement of shorter programs. A minimum block overlap of 66% (i.e., minimum 2 s of overlap) between consecutive analysis windows is required. By doing this, a short but very loud event will not affect the Loudness Range of a longer segment. Similarly, the fade-out at the end of a music track, for example, will not increase the Loudness Range noticeably. Specifically, the range of the distribution of loudness levels is determined by estimating the difference between a low and a high percentile of the distribution.

The Loudness Range also employs a gating method. The relative threshold is set to a level of -20 LU relative to the absolute gated loudness level. Certain types of programs may be, overall, very consistent in loudness, but have some sections with very low loudness, such as only the background environment. If the Loudness Range did not use gating, programs like that would (incorrectly) get quite a high Loudness Range measurement.

The purpose of the absolute threshold gate is to make the conversion from the relative threshold to an absolute level robust against longer periods of silence or low-level background noise. The absolute threshold is set to -70 LUFS, because no relevant signals are generally found below this loudness level.

The optimum LRA will be program dependent.

Maximum True Peak Level

The maximum true peak level is defined to -1 dBTP measured with a meter compliant with both ITU-R BS.1770 and EBU Technical Document 3341. Hence four times oversampling is applied to ensure a correct reading.

EBU Mode Meter

To ensure that the different descriptors are measured and reported correctly and not mixed up with other measures, any meter that measures according to R 128 must have an "EBU Mode." When set in this mode the meter complies with EBU Technical Document 3341. The EBU Mode does not concern the graphical/UI details or the implementation of a meter.

THE THREE TIME SCALES

There are three time scales, represented as follows:

- The shortest time scale is called "momentary," abbreviated "**M**."
- The intermediate time scale is called "short-term," abbreviated "**S**."

- The program- or segment-wise time scale is called "integrated," abbreviated "**I**."

The loudness meter should be able to display the maximum value of the "momentary loudness." This maximum value is reset when the integrated loudness measurement is reset.

INTEGRATION TIMES AND METHODS, METER BALLISTICS

In all cases the loudness measurement is performed as specified in ITU-R BS.1770. The measurement parameters for EBU Mode are the following:

- The **momentary loudness** measurement uses a sliding rectangular time window of length 0.4 s. The measurement is not gated.
- The **short-term loudness** measurement uses a sliding rectangular time window of length 3 s. The measurement is not gated. The update rate for "live meters" shall be at least 10 Hz.
- The **integrated loudness** measurement uses gating as described in ITU-R BS.1770. The update rate for live meters shall be at least 1 Hz.

The EBU Mode loudness meter provides functionality that enables the user to at least start/pause/continue the measurement of integrated loudness and Loudness Range simultaneously, that is, switch the meter between "running" and "standby" states; and reset the measurement of integrated loudness and Loudness Range simultaneously, regardless of whether the meter is in the "running" or "standby" state.

THE MEASUREMENT GATE

The "integrated loudness" should be measured using the gating function described earlier.

LOUDNESS RANGE (LRA) DESCRIPTOR

An EBU Mode meter should compute the Loudness Range (see above).

UNITS

A **relative** measurement, such as relative to a reference level or a range: $L_K = $ xx.x LU

An **absolute** measurement: $L_K = $ xx.x LUFS

The "L" in "L_K" indicates loudness level; the "K" indicates the frequency weighting used.

TRUE PEAK MEASUREMENT

True peak is measured using four times oversampling.

SCALES AND RANGES

The display of an EBU Mode meter may be simply numerical, or an indication on a scale.

The scale used may either be an absolute scale, using the LUFS unit, or alternatively the zero point may be mapped to some other value, such as the

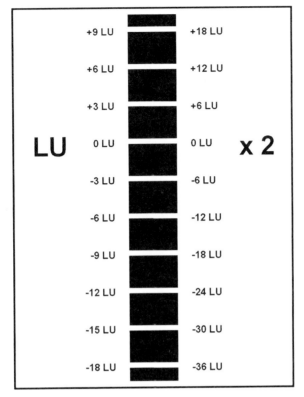

FIGURE 14.7 EBU Mode scales. The recommendation R 128 does not specify any specific layout. However, if in the EBU mode, these scales should be available.

target loudness level (as in ITU-R BS.1771). In the latter case the LU unit should be used, indicating a relative scale. For an EBU Mode meter, the target loudness level shall be −23 LUFS = 0 LU (as defined in EBU R 128). An EBU Mode meter should offer both the relative and the absolute scale.

The location of the target/reference loudness level should remain the same, regardless of whether an absolute or relative scale is displayed.

An EBU Mode meter should offer two scales, selectable by the user:

- Range −18.0 LU to +9.0 LU (−41.0 LUFS to −14.0 LUFS), named the "EBU +9 scale"
- Range −36.0 LU to +18.0 LU (−59.0 LUFS to −5.0 LUFS), named the "EBU +18 scale"

The "EBU +9 scale" should be used by default.

Bibliography

EBU Technical Recommendation R 128 Loudness normalisation and permitted maximum level of audio signals (2010).

EBU Technical Document 3341, Loudness Metering. 'EBU Mode' metering to supplement Loudness normalisation according to EBU Technical Document R 128 (2010).

EBU Technical Document 3342 (2010), Loudness Range. A descriptor to supplement Loudness normalization according to EBU Technical Recommendation R 128.

Itu-R BS. 1770. Algorithms to measure audio programme loudness and true peak audio level (2006).

Itu-R BS. 1771. Requirements for loudness and true-peak indicating meters (2006).

TC Electronic: Manual for LM5 & LM5D Loudness Radar Meters (2010).

Chapter | fifteen

Calibration of Level Meters

Attention must be paid to the fact that a level meter is a measuring instrument. And if a measuring instrument is not calibrated, what can it then be used to measure?

CONSTANT TONE

An analog instrument is calibrated first and foremost using a constant pure tone. In many cases the frequency is chosen to be 1 kHz. However, that will depend on the type of meter or instrument you want to calibrate. Sound level meters or any measuring device related to acoustic measurements can be calibrated using a constant continuous 1 kHz pure tone because at this frequency the weighting filters (like IEC A) are neutral. Many program meters will accept or require 1 kHz as well for checking the constant continuous reading. In loudness meters (ITU/EBU) one should be aware that 1 kHz actually lies on the slope of the pre-filter. Hence very narrow tolerances are needed in the filter design.

If you are in possession of a high precision digital RMS voltmeter, you can then measure the magnitude of the signal for the comparison.

TONE BURST

The integration time — or perhaps more correctly — the reaction time of an instrument is tested using a tone burst generator. The practical calibration is usually defined by the time it takes the meter to reach a given percentage of the reference level. A tone burst generator can supply a tone for a well-defined period of time, for example 5 ms or 300 ms, and repeat at a fixed interval, for example 1.5 s or 300 ms. It should be noted that the frequency generally is

Audio Metering. DOI: 10.1016/B978-0-240-81467-4.10015-2

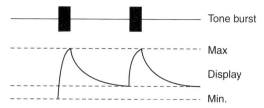

FIGURE 15.1 Testing an instrument's integration time using a tone burst. The individual standard specifies what the instrument should display within given time intervals.

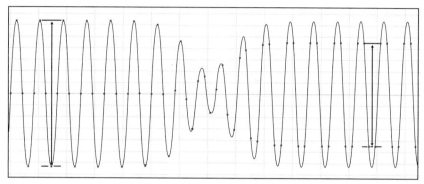

FIGURE 15.2 A crossfade between two 12 kHz sine waves, both generated to have the same level, and at a sampling frequency of 48 kHz, which provides four samples per period. On the left the tone generated has its first sample at 0 degrees (the dots), and hence the others at 90, 180, and 270 degrees of the sine wave period. On the right the first sample was made at 45 degrees and the following samples at 135, 225, and 315, respectively. The peak level of the first part is calculated by a digital meter to be 3 dB higher than that of the second part.

5 kHz in order to have a sufficient number of periods to comprise a tone burst. If the frequency is too low or the burst does not contain an integer number of periods there is a risk of measuring just a limited portion of a single period, which leaves the test signal with frequency components of higher order. IEC 60268-10 requires 10 kHz for very short impulses as at least five periods are required to form a suitable tone burst.

Test signals for the calibration of instrumentation can also be found on various CDs. However, if used for analog inputs or analog instruments, the analog output of the CD players must be checked. Using a computer-based editing program, it is possible to generate customized test tone sequences.

Test signals for digital instruments must be designed in a way that does not compromise the sampling rate. For instance, if the test frequency is exactly one fourth of the sampling rate there may be a problem.

PROCEDURE

The analog instrument is first calibrated with a constant tone at the alignment level (reference indication). Then the tone burst signal is applied, with the individual tone "packages" having the same level as the constant tone.

SVI or VU

According to IEC 60268-17 a Standard Volume Indicator or VU instrument must reach 99% (±10 %) of the reference indication (the "0" marking) for a tone length of 300 ms. If the tone length is shorter, then the reading must be lower. The fallback time should be identical to the reaction time, meaning the reading should fall to the bottom of the scale in 300 ms.

PPM (QPPM)

The reference indication of the IEC PPM (Quasi Peak Program Meter) instrument must be as follows according to IEC 60268-10:

Type I	1.55 V (+6 dB on Nordic scale)
Type IIa	1.94 V (6 on the BBC scale)
Type IIb	2.18 V (+9 on the IIb scale)

When using the instruments to check line levels the accuracy of the scales must be checked as well by supplying other levels.

The dynamic response is defined by the reaction time. The Type I PPM must reach a level 2 dB below the reference indication for a tone length of 5 ms. In the pause of 1.5 seconds, the instrument must manage to fall 20 dB. The reading of the meter should be in accordance with the following scheme:

TABLE 15.1 Dynamic response for IEC PPM instruments in normal mode. However, for a tone burst with a duration of 5 ms, the display must reach 2 dB from the display that occurs at a constant tone.

Burst, duration	ms	Display dB	Tolerance dB
IEC Type I	10	−1	±0.50
	5 (ITU recommendation)	−2	±1.00
	3	−4	±1.00
	0.4	−15	±4.00
IEC Type IIa	100	6	±0.50
	10	5.5	±0.50
	5	5	±0.75
	1.5	3.75	±1.00
	0.5	1.75	±2.00
IEC Type IIb	100	+8	±0.50
	10	+6	±0.50
	5	+4	±0.75
	1.5	−1	±1.00
	0.5	−9	±2.00

LOUDNESS METER, EBU MODE

Calibrating the loudness meter is a little tricky, as there are several parameters to check. The meter will read both linear audio for the TP (True Peak) and k-weighted audio for the loudness reading resulting in LUFS (Loudness Unit with reference to full scale). The loudness reading includes gating for the integrated loudness. Additionally, there is the LRA (loudness range).

In the EBU mode these k-weighted readings are available: M (momentary loudness), S (short-term loudness) and I (integrated loudness).

Generally, the ITU prepares the test procedures for metering in broadcast. However, these procedures have not been finished yet. In the meantime the EBU has recommended the following series of test signals for alignment and calibration:

TABLE 15.2 Minimum requirements for test signals used with the EBU loudness meter.

Test case	Test signal	Expected response and accepted tolerances
1	Stereo sine wave, 1000 Hz, −23.0 dBFS (per-channel peak level); signal applied in phase to both channels simultaneously; 20 s duration	M, S, I = −23.0 +/−0.1 LUFS M, S, I = 0.0 +/−0.1 LU
2	As #1 at −33.0 dBFS	M, S, I = −33.0 +/−0.1 LUFS M, S, I = −10.0 +/−0.1 LU
3	As #1, preceded by 20 s of −40 dBFS stereo sine wave, and followed by 20 s of −40 dBFS stereo sine wave	I = −23.0 +/−0.1 LUFS I = 0.0 +/−0.1 LU
4	As #3, preceded by 20 s of −75 dBFS stereo sine wave, and followed by 20 s of −75 dBFS stereo sine wave	I = −23.0 +/−0.1 LUFS I = 0.0 +/−0.1 LU
5	As #3, but with the levels of the 3 tones at −26 dBFS, −20 dBFS, and −26 dBFS, respectively	I = −23.0 +/−0.1 LUFS I = 0.0 +/−0.1 LU
6	5.0 channel sine wave, 1000 Hz, 20 s duration, with per-channel peak levels as follows: −28.0 dBFS in L and R −24.0 dBFS in C −30.0 dBFS in Ls and Rs	I = −23.0 +/−0.1 LUFS I = 0.0 +/−0.1 LU
7	Authentic program 1, stereo, narrow loudness range (NLR) program segment; similar in genre to a commercial/promo	I = −23.0 +/−0.1 LUFS I = 0.0 +/−0.1 LU
8	Authentic program 2, stereo, wide loudness range (WLR) program segment; similar in genre to a movie/drama	I = −23.0 +/−0.1 LUFS I = 0.0 +/−0.1 LU

Chapter sixteen

Relationships Between Scales

CHAPTER OUTLINE

When it comes to level meters, the abundance and different types of appearances of the scales are overwhelming. However, most of them share a common trait in that their scales use divisions that are based on units in dB. Yet the portion of the dynamic range the scales cover can be different. For certain purposes, it is important to be able to monitor the entire dynamic range. In other cases, it is only important to see what is occurring in the immediate vicinity of full modulation. Some scales combine both properties.

As far as professional equipment is concerned, good and readable scales predominate. With consumer equipment, it is often the case that the aim is to just have something that moves, and here it is rare that there is any real possibility of performing a calibration.

COMMENTS ON THE SCALES

A number of the scales currently used are shown in Figure 16.1 together with their relationship. Some comments follow in connection with how the scales are shown.

Volts

The scale in volts is included for the sake of comparison. This of course is the RMS value.

Audio Metering. DOI: 10.1016/B978-0-240-81467-4.10016-4

volts	dBu	IEC I Nordic	IEC IIa BBC	IEC IIb	DIN	SVI or VU*) Direct reading	North Am., Australia	France	EBU AD/DA
12.28	24								
10.95	23								
9.76	22								
8.70	21								
7.75	20								
6.91	19								
6.16	18								0
5.49	17								-1
4.89	16								-2
4.36	15								-3
3.88	14								-4
3.46	13								-5
3.09	12	12	7	12					-6
2.75	11	11		11	5				-7
2.45	10	10		10	4				-8
2.18	9	9		9	3				-9
1.95	8	8	6	8	2				-10
1.74	7	7		7	1	+3			-11
1.55	6	6		6	0	+2			-12
1.38	5	5		5	-1	+1			-13
1.23	4	4	5	4	-2	0			-14
1.10	3	3		3	-3	-1	+3		-15
0.976	2	2		2	-4	-2	+2		-16
0.870	1	1		1	-5	-3	+1	+3	-17
0.775	0	test	4	test	-6	-4	0	+2	-18
0.691	-1	-1		-1	-7	-5	-1	+1	-19
0.616	-2	-2		-2	-8	-6	-2	0	-20
0.549	-3	-3		-3	test	-7	-3	-1	-21
0.489	-4	-4	3	-4	-10	-8	-4	-2	-22
0.436	-5	-5		-5	-11	-9	-5	-3	-23
0.388	-6	-6		-6	-12	-10	-6	-4	-24
0.346	-7	-7		-7	-13	-11	-7	-5	-25
0.309	-8	-8	2	-8	-14	-12	-8	-6	-26
0.275	-9	-9		-9	-15	-13	-9	-7	-27
0.245	-10	-10		-10	-16	-14	-10	-8	-28
0.218	-11	-11		-11	-17	-15	-11	-9	-29
0.195	-12	-12	1	-12	-18	-16	-12	-10	-30
0.174	-13	-13			-19	-17	-13	-11	-31
0.155	-14	-14			-20	-18	-14	-12	-32
0.138	-15	-15			-21	-19	-15	-13	-33
0.123	-16	-16			-22	-20	-16	-14	-34
0.109	-17	-17			-23		-17	-15	-35
98 m	-18	-18			-24		-18	-16	-36
87 m	-19	-19			-25		-19	-17	-37
78 m	-20	-20			-26		-20	-18	-38
69 m	-21	-21			-27			-19	-39
61 m	-22	-22			-28			-20	-40
55 m	-23	-23			-29				-41
49 m	-24	-24			-30				-42
44 m	-25	-25			-31				-43
39 m	-26	-26			-32				-44
35 m	-27	-27			-33				-45
31 m	-28	-28			-34				-46
28 m	-29	-29			-35				-47
⇓	⇓	⇓	⇓	⇓	⇓				⇓
-	-	-42	-∞	-∞	-50				-

FIGURE 16.1 Relation between common scales*) see the section. SVI or VU.

dBu

The unit dBu is an absolute magnitude with the reference 0.775 V. Most of the scales express the voltage of the signal.

IEC I, Nordic Scale

This scale is the most important among professionals and broadcasters in the Nordic countries.

IEC IIa, BBC

With respect to the standard, there are 4 dB between each subdivision. However, there are some instruments with 6 dB between the lowest steps on the scale.

IEC IIb

This scale is used for the adjustment of transmission lines.

DIN

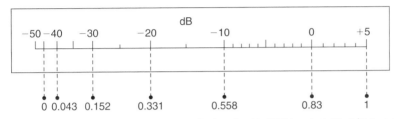

FIGURE 16.2 Scale subdivisions as originally described in "Pflichtenheft Nr. 3/6, Inst. für Rundfunktechnik" (Requirements for Broadcasting).

SVI or VU

This scale is used with little adherence to any standard. It has to be mentioned that the scale originally did not have VU as a unit. The VU on the scale only indicates that this instrument is a Standard Volume Indicator. The reading of the instrument is found on the attenuator in front of the meter.

That said, we must acknowledge the different ways this instrument and its scale is used — mostly without applying the attenuator.

There are three different scale offsets shown in Figure 16.1:

Direct reading: When 1.23 V is applied, the reading on the scale is 0 if either no attenuator is used in front of the meter — or the attenuator in front of the meter is set in a position where it does not attenuate. However, the volume is actually +4 (dBm) according to the standard.

In North America and Australia, this is actually the correct reading if no attenuation is applied. In France, the scale is further displaced so that 0 on the scale corresponds to 0.616 V.

M/S Scales

Some meters provide an M/S scale, displaying the direct relation of the in-phase contents compared to the anti-phase contents of the left and right channel.

Loudness Scales

The loudness scales cannot be compared unless the algorithms forming the reading are exactly identical.

EBU vs. SMPTE

The last scale shows the relation to EBU R68, which gives the correlation between the analog signal level and its digital coding. Yet another standard exists: SMPTE RP155. Unfortunately, they are not identical. The relationship between the two is shown in Figure 16.1. It shows that 0 dBFS corresponds to

EBU R68			SMPTE RP155			
dBFS	dBu	Volts	dBFS	dBu	Volts	
0	18	6.16	0	24	12.28	**Max level**
-1	17	5.49	-1	23	10.95	
-2	16	4.89	-2	22	9.76	
-3	15	4.36	-3	21	8.70	
-4	14	3.88	-4	20	7.75	
-5	13	3.46	-5	19	6.91	
-6	12	3.09	-6	18	6.16	
-7	11	2.75	-7	17	5.49	
-8	10	2.45	-8	16	4.89	
-9	9	2.18	-9	15	4.36	
-10	8	1.95	-10	14	3.88	
-11	7	1.74	-11	13	3.46	
-12	6	1.55	-12	12	3.09	
-13	5	1.38	-13	11	2.75	
-14	4	1.23	-14	10	2.45	
-15	3	1.09	-15	9	2.18	
-16	2	0.976	-16	8	1.95	
-17	1	0.870	-17	7	1.74	
-18	0	0.775	-18	6	1.55	**Test**
-19	-1	0.691	-19	5	1.38	
-20	-2	0.616	-20	4	1.23	**0 VU**
-21	-3	0.549	-21	3	1.09	
-22	-4	0.489	-22	2	0.976	
-23	-5	0.436	-23	1	0.870	
-24	-6	0.388	-24	0	0.775	
-25	-7	0.346	-25	-1	0.691	
-26	-8	0.309	-26	-2	0.616	
-27	-9	0.275	-27	-3	0.549	
-28	-10	0.245	-28	-4	0.489	
-29	-11	0.218	-29	-5	0.436	
-30	-12	0.195	-30	-6	0.388	
-31	-13	0.174	-31	-7	0.346	
-32	-14	0.155	-32	-8	0.309	
-33	-15	0.138	-33	-9	0.275	
-34	-16	0.123	-34	-10	0.245	
-35	-17	0.109	-35	-11	0.218	
-36	-18	0.098	-36	-12	0.195	
-37	-19	0.087	-37	-13	0.174	
-38	-20	0.078	-38	-14	0.155	
-39	-21	0.069	-39	-15	0.138	
-40	-22	0.062	-40	-16	0.123	
-41	-23	0.055	-41	-17	0.109	
-42	-24	0.049	-42	-18	0.098	

FIGURE 16.3 Relationship between the analog signal level and digital coding in converters as per the EBU and SMPTE standards, respectively.

+18 dBu according to the EBU standard. Similarly, 0 dBFS is equal to +24 dBu according to SMPTE. Hence, you must know the conversion factor in order to ensure that inexplicable jumps in level of up to 6 dB do not occur. This will happen in particular if you have a mixture of American and European sources.

Chapter | seventeen

Dynamic Scales

CHAPTER OUTLINE

It is a fact that the dynamic range can vary in different types of program material. Pop music can be very compressed and thus have a very limited dynamic range. Classical music and films mixed for the cinema have a large dynamic range. When recording, this means that the need for headroom will vary depending on the program material.

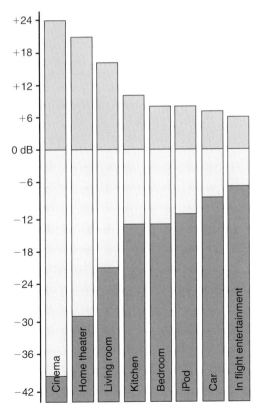

FIGURE 17.1 Dynamic ranges in various environments normalized with reference to dialog level. (Ref. Thomas Lund, TC Electronic.)

Audio Metering. DOI: 10.1016/B978-0-240-81467-4.10017-6

The listening location also affects the dynamic range that is necessary. If the material is being listened to in an automobile, then the dynamics must be limited in order for the audio to be heard above the background noise. If the material is to be played back in the home where it must not be too loud, there is similarly a need for limited range.

Over the years proposals have been advanced by a number of parties for implementing dynamic scales where it would be possible to utilize the dynamics of the storage media or the transmission line optimally. An applicable factory standard from Dolby will be described in the following, as well as a proposal from an American producer.

DTV

In the US, ATSC (Advanced Television Systems Committee) was introduced as a standard for DTV (Digital TV). In this system, it is possible to reproduce up to six channels of sound encoded in Dolby Digital.

Along with the digitized (and bit-reduced) sound, supplemental information is sent in the form of metadata. In popular terms it can be said that the information previously written on the box containing the tape is now included in the signal itself. What is interesting about the information that has been included as metadata is that, among other things, it can contain different possibilities for level, dynamic range, number of channels, etc., which can then be selected when the program is listened to.

The bit stream that contains the digital sound is divided into frames and blocks. The metadata can either be attached to each frame or to each block (compression data for example). Other metadata applies to the entire program (for example, information on level). It is the sender who decides which parameters will be associated with the program. However, the user can select from the possibilities offered and customize the sound per their requirements.

DIALNORM

One of the biggest problems with TV sound is the difference in level from program to program and station to station. When a viewer flicks through channels, the levels encountered are quite different, particularly with respect to speech. It is in a way quite natural, when considered in relation to full modulation, for the speech level (dialog) in a film to be at a lower level than the speech level in an interview program. One can of course decide always to have speech at a specific (low) level; however, that would mean a poor utilization of the system's dynamic range.

Instead, "Dialnorm" – an abbreviation of dialog normalization – has been introduced. This lets the program material be recorded in an optimum manner, but the level at which the dialog is situated is specified. All the audio can subsequently be shifted in level so that the volume of the dialog is uniform from program to program.

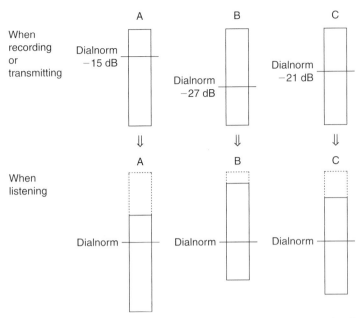

FIGURE 17.2 Dialnorm is used to adjust dialog (or the program's average level) to the same listening level, regardless of whether the individual programs were recorded with different level placements, headrooms and dynamic ranges.

The level at which the dialog is played back is also determined by the selected dynamic range. If the program is played back uncompressed (setting: "lin"), then the dialog level will lie at −31 dBFS. If the program is played back with high compression (setting: "RF"), then the dialog level will lie at −20 dBFS.

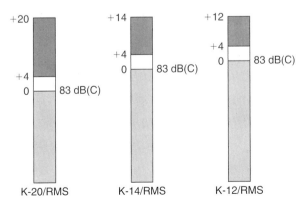

FIGURE 17.3 The K system. The three scales are specified here with an RMS detector. The scales can also be used in connection with the display of L_{eq} and Loudness. [*] X can be 20, 14, or 12.

THE K SYSTEM

In an attempt to better utilize the dynamic range when recording and to have an indication of the modulation and subjective loudness, a proposal has been presented for a meter system by Bob Katz, an American whose experience includes many years of recording and mastering in the American music industry.

The K system — named after him — builds upon three scales, namely K-20 with 20 dB headroom over 0 dB, K-14 with 14 dB, and K-12 with 12 dB headroom. On the scales, the color green is used below 0 dB, yellow between 0 and +4 dB, and red for over +4 dB. Each scale can be used with three different time/frequency weightings, named RMS, LEQA, and Zwicker.

Thus K-X[*)]/RMS is used with a frequency response of 20 Hz — 20 kHz ±0.1 dB.

K-X[*)]/LEQA uses A-weighting (IEC A) and an integration time of 3 seconds.

K-X[*)]/Zwicker uses, as the name suggests, Zwicker's model for loudness.
[*)]X can be 20,14, or 12.

For calibrated playback, pink noise at a signal level of 0 dB on the scale will correspond to 83 dB(C) in the listening position.

Note that the recording scale goes together with an acoustic listening/monitoring level specified in dB(C).

Polarity and Phase Reading

CHAPTER OUTLINE

In the production, recording, and transmission of stereo signals, it is not just the levels of the signals in relation to each other that need to be monitored, but also the polarity or phase relationship between them. This chapter describes some important instruments that are used for this purpose.

POLARITY

Polarity is a question of the audio signals being "in phase" or "out of phase," meaning the positive parts of the waveform become negative and the negative becomes positive. If only looking at — or listening to — one channel (mono) audio being 180° out of phase, it is not always recognized as a problem. However, in professional audio no phase inversion should be accepted if not done on purpose.

Looking at the waveform of two channels the "out of phase" situation corresponds to only one of the two channels being inverted.

In analog audio this polarity problem is an error far more common than one would imagine. In loudspeaker manufacturing a very high proportion of units are actually born with reversed polarity by inadvertently swapping the voice coil leads on the soldering terminals. It is easy to go wrong in the soldering of connectors, patch bays, etc. Most power amplifiers retain polarity of the signal from input to output; however, others do not. So just having different amp brands powering the speaker systems may cause problems.

All equipment should retain the polarity from input to output. To check this, a number of devices of varying levels of sophistication are available. Basically a DC impulse (an impulse only deflecting in only one direction) is generated acoustically when including a microphone in the audio chain or electrically for the microphone or line inputs. The receiver detects the direction of the impulse

Audio Metering. DOI: 10.1016/B978-0-240-81467-4.10018-8

electrically on the output or acoustically if the loudspeaker is included in the audio chain. The readout is basically an LED display indicating whether the received signal is in phase or out of phase when compared with the transmitted signal.

Please note that in some two-way or three-way loudspeaker systems the midrange speaker may be measured as being out of phase. This is due to the crossover technique if using 2nd order (12 dB/octave) filters. At the crossover frequency the phaseshift is +90° in one filter and −90° in the other. In order to make the loudspeaker units work together at this point the signal to one of the speakers is reversed.

THE PHASE METER

This instrument is used for stereo recording and monitoring. It has a scale ranging from "+1" to "−1" to show the current phase difference between the signals in the left and right channels. Instead of giving the phase angle in degrees, it displays the cosine of the phase angle (the phase difference).

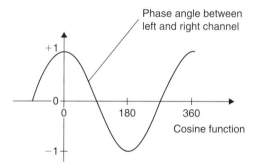

FIGURE 18.1 The cosine function is shown here. When the phase angle between left and right is 0° the cosine is +1. When the phase angle is 180° the cosine is −1.

The cosine of an angle of 0° is +1.
The cosine of 90° is 0.
The cosine of 180° is −1.
If the left and right signals are completely in phase (i.e., the phase angle is 0°), then the instrument indicates "1."
If the signals are completely out of phase (i.e. the phase angle is 180°), then the instrument indicates "−1."
If the phase angle is 90°, the instrument indicates "0."
If there is only a signal in one of the channels, then the instrument will also display "0" since this position is the neutral position.
When stereo signals are being recorded, the indication should normally lie between "0" and "1."

Integration Time

Normally, the instrument is equipped with a relatively long integration or averaging time, typically 600 ms; this means that a slow indication is given, but it is

FIGURE 18.2 The scale of the correlation or phase meter is shown above. It has values from +1 to −1.

still fast enough to show the phase relationship of varying signals at low frequencies.

Phase meters are normally used to assess recordings for gramophone records, where reversed phase at lower frequencies can be synonymous with the stylus just about having to leave the surface of the record if no compensation is performed. The instrument is, however, also used in all types of multi-channel productions — in particular, for the production of multimedia sound files, especially when the audio is played back through matrix systems that generate multi-channel audio on the basis of two channels. When the instrument reads values below 0 the audio disappears into the rear channel.

GONIOMETER — AUDIO VECTOR OSCILLOSCOPE

The goniometer or "audio vector oscilloscope" is an instrument that can give a detailed picture of the relationships in a stereo signal — or between two arbitrary signals. The idea behind the instrument was developed by Holger Lauridsen, the Chief Engineer at Danmarks Radio in the 1940s and early 1950s. It involves an oscilloscope with one input being used for the X-deflection and the other input for the Y-deflection.

In comparison with a normal oscilloscope, the image is rotated by 45°. This means that, when the same signal is applied in the left and right channels (X and Y) it leads to a vertical deflection (a vertical line); if the signals are otherwise identical but in opposite phase, then a deflection will occur in the horizontal plane (horizontal line). When the signals are different, the display changes from straight lines to spatial figures. What is ingenious about the instrument is that it can show many different parameters in the stereo signal simultaneously.

As long as the major portion of the display lies within ±45° in relation to the vertical axis, then there is a high degree of mono compatibility. When the curves begin to bend at the top and bottom, it can indicate nonlinearity in the relationship between the two channels. This can arise if overloading or limiting is occurring in only one channel. It can stem from time delays between the channels that may arise in older digital systems or tape units.

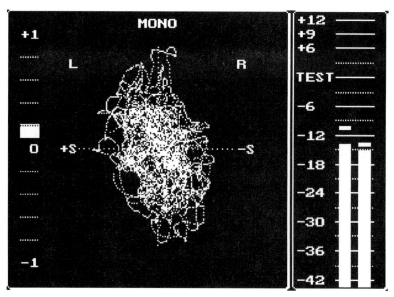

FIGURE 18.3 The stereo signal is wide, but there is still a high degree of mono compatibility. Regarded as optimum.

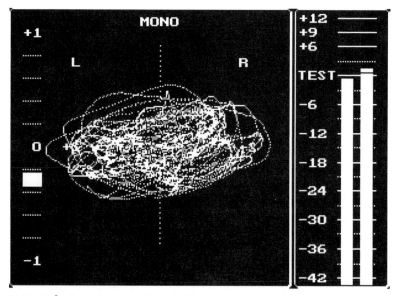

FIGURE 18.4 The stereo signal is too wide. The mono compatibility is small, i.e., some of the signal will disappear in mono.

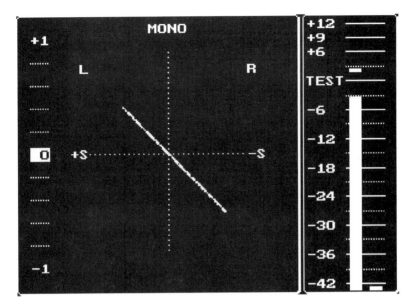

FIGURE 18.5 Signal only in the left channel.

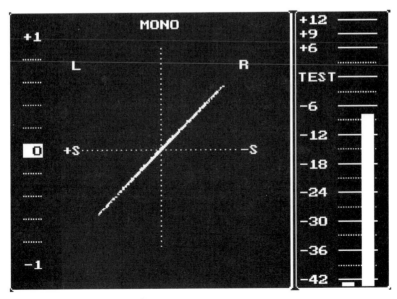

FIGURE 18.6 Signal only in the right channel.

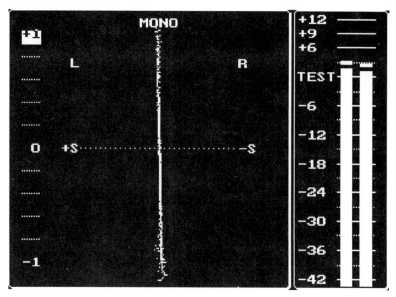

FIGURE 18.7 Almost a pure mono signal in the two channels.

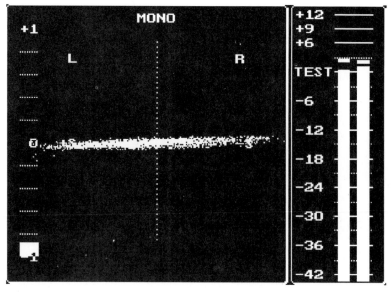

FIGURE 18.8 Almost a pure mono signal, but in opposite phases in the two channels.

SPECTRAL PHASE DISPLAY

This display is developed as a special analysis tool for digital audio workstations. It shows the phase difference, in degrees, between the left and the right channels recorded. Due to the various display colors frequency selective readings are possible. Hence it is possible to distinguish the phase of complex two-channel signals. The horizontal axis is time. The vertical axis is the phase (+/− 180°), with 0° in the middle. As mentioned, the colors can be defined for different ranges of the frequency spectrum. See the Figures 18.3−18.8 for practical applications of this display.

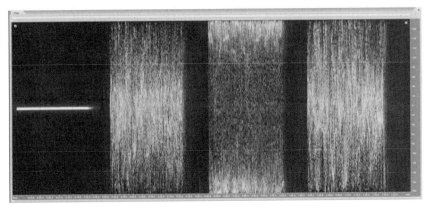

FIGURE 18.9 This spectral phase display show three sets of microphones recording an organ in a small 12th century church. The first part is a mono (omnidirectional) microphone, the second part shows the phase distribution of a MS setup placed further away. The third microphone setup is a reversed XY setup. The fourth part show all three setups recorded simultaneously. This was created in Adobe Audition 3.

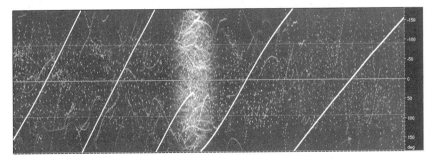

FIGURE 18.10 In this case the display is used for phase comparison between an extracted ENF-component (electric network frequency) and a fixed reference. This analysis is performed for authentication of digital recordings in the field of audio forensics; this was created in Adobe Audition 3.

Chapter nineteen

Display of Level Distribution

CHAPTER OUTLINE

In a number of contexts it is interesting to be able to see what the signal level has been over a longer period of time. This can be done either by plotting the level along a time axis or by performing a statistical analysis that shows the level distribution over a time period.

LEVEL RECORDER

The level recorder is an instrument that has been used for nearly all forms of sound analysis. The recorder can either be mechanical with paper and pen or it can be a part of a computerized system, where data is collected on an ongoing basis and printed out afterwards.

The recorder must have an integration time. It can either be the mechanical properties of the recorder that are the crucial factor or it can be the integration time of the level meter that is used.

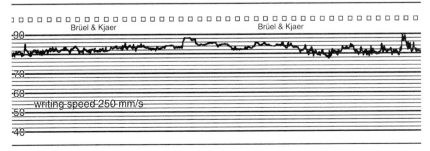

FIGURE 19.1 A traditional recording strip, where the resolution is determined by the speed at which the pen writes. (Recorder: Brüel & Kjær)

Audio Metering. DOI: 10.1016/B978-0-240-81467-4.10019-X

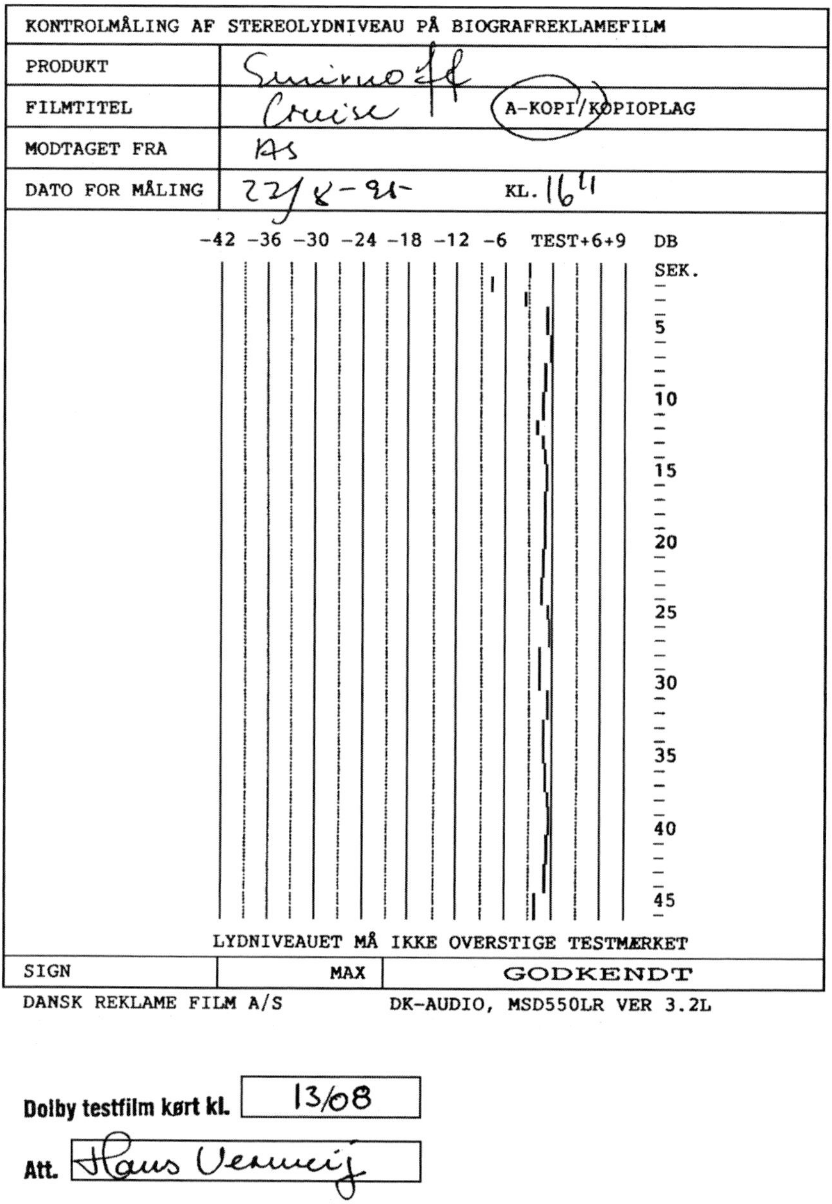

KONTROLMÅLING AF STEREOLYDNIVEAU PÅ BIOGRAFREKLAMEFILM

PRODUKT	Smirnoff	
FILMTITEL	Cruise	A-KOPI/KOPIOPLAG
MODTAGET FRA	AS	
DATO FOR MÅLING	22/8-91	KL. 16[41]

-42 -36 -30 -24 -18 -12 -6 TEST+6+9 DB

SEK. 5 10 15 20 25 30 35 40 45

LYDNIVEAUET MÅ IKKE OVERSTIGE TESTMÆRKET

SIGN	MAX	GODKENDT

DANSK REKLAME FILM A/S DK-AUDIO, MSD550LR VER 3.2L

Dolby testfilm kørt kl. 13/08

Att. Hans Vermeij

FIGURE 19.2 Example of a level recording for checking and documentation of the sound level of an advertisement. The requirement is that the level must not exceed "TEST" on the Nordic Scale, which is equal to 0 dBu.

HISTOGRAM

The histogram is a function that can show the statistical distribution within the given time interval under consideration.

Used in connection with level meters, the figure plotted on one axis may express the signal levels that have occurred in the time interval. On one axis

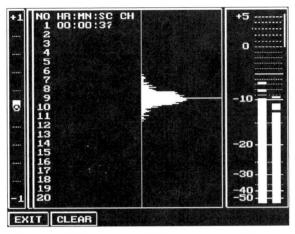

FIGURE 19.3 Example of a histogram in which it can be seen that the level of a recording has been at around −10 dB for a large portion of the time.

the figure will show the frequency at which the signal occurs within all the individual levels. This is shown by columns, where the length of the individual column represents that portion of the time period in which the level concerned occurs. For the sake of clarity, the column for the level that occurs often can be used to normalize the display, i.e., this column will be given the fullest possible length of that display.

Among other things, the histogram can be used to show how the signal levels are distributed in an audio sequence, for example in a piece of recorded music. Information on the dynamic range of the recording can be obtained in this way. In signals with a large degree of dynamic range such as classical music, the plot will extend across a large portion of the vertical axis, which indicates a large scatter in the levels.

If a lot of compression is used, as for example in pop music, a very narrow figure of only a limited extent in the vertical direction will be traced out. For purposes of comparison, it should be mentioned that a constant tone will only result in a single horizontal column, next to the level concerned.

In histograms for digital recordings of pop music, it is possible to observe that the level exceeds 0 dBFS, which is not actually possible. However, this is due to the fact that a calculation is being performed of the level. It is presumed that when more than one sample exceeds the maximum, then it must involve overloading.

OVERLOAD DISPLAY

It is not always possible to sit and keep an eye on the instrument when a recording is being checked for possible overloading. It is therefore practical to have a function that can log the points in time at which overloading occurs during a recording.

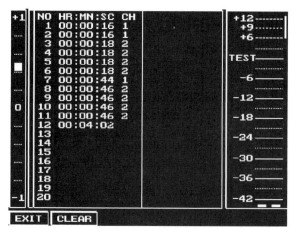

FIGURE 19.4 An example is shown here of a listing of overload in a stereo recording. The point in time at which the overloading occurs, as well as the channel, is shown.

As mentioned previously in the Histogram section, overloading is not found in the digital world even when people speak of levels beyond 0 dBFS.

Clipping of a signal can be registered as multiple samples that lie successively at the level of 0 dBFS. By a counting process it is therefore possible to

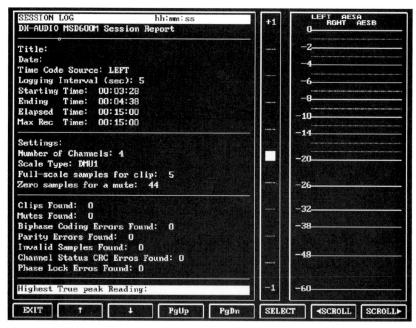

FIGURE 19.5 An example is shown here of a report covering a session of 15 s. Notice that the definition for overloading — and hence for clipping — has been set to 5 full-scale samples.

discern when the digital signal has been overloaded. In certain systems it is possible to set how many 0 dBFS samples must occur before it is registered as an overload.

It is possible to log the overload condition using a stopwatch technique or timecode.

CUMULATIVE DISTRIBUTION

In connection with noise measurements, a statistical analysis can be used that gives the cumulative distribution of the sound levels recorded. In the graphical rendering, a curve is shown where one axis is the A-weighted sound pressure level, while the other axis represents the percentage share of the time interval under consideration. The curve then expresses what sound pressure level has been exceeded for what part of the time. A concept such as background noise in the external environment is typically defined as the A-weighted sound pressure level that is exceeded during 95% of the time period.

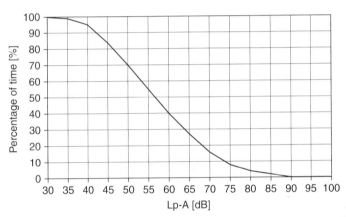

FIGURE 19.6 Example of cumulative distribution. The curve shows for what part of the time interval under consideration a given level is exceeded. For example, it can be seen that the level of 60 dB(A) is exceeded 40% of the time.

Chapter twenty

Multi-Channel and Surround Sound

CHAPTER OUTLINE

Audio Metering. DOI: 10.1016/B978-0-240-81467-4.10020-6

Audio using more than two channels is standard in many areas, such as audio productions for TV, the cinema, the home cinema, the car, and music formats for Hi-Res CD, DVD, Blu-Ray, and audio formats for computer-based games.

These formats involve sound using four to eight channels. Certain formats are stored and transmitted in just two channels, and are "unpacked" when they are listened to. Other formats are produced as discrete channels, but are stored and transmitted digitally as a single bit stream. Furthermore, concepts such as "virtual surround" exist where only two channels are used and these channels give the impression that there is sound all the way around the listener.

In order to know what means are available for monitoring signals, the way in which most important formats work will be described here. Emphasis will be placed on the matrix-encoded formats and their principles, due to the fact that these are quite widespread in the multimedia area. We will also look at the relation between the signal levels on tape, film, and hard disk, and look at the associated acoustic sound levels.

MATRIX-ENCODED FORMATS

The matrix-encoded formats are characterized by four or five channels being mixed down to two. A network mixes the signals in a manner that permits their separation again although this is not possible under all conditions.

The principle is also called 4:2:4 or 5:2:5, which indicates how many channels are produced, how many are stored or transmitted and finally how many channels the format is reproduced in. The most widespread systems are Dolby Stereo (or Dolby Surround), which is a 4:2:4 system. Others include Circle Surround, 5:2:5, and Lexicon Surround, 4:2:5 or 5:2:5.

DOLBY® STEREO & DOLBY® SURROUND

Dolby Stereo and Dolby Surround are two sides of the same coin. Dolby Stereo is the original name for the matrix-encoded audio signals found on the two optical sound tracks of cinema films. Here, noise reduction is also used, either Dolby A or Dolby SR. Dolby Surround uses the same matrix system as Dolby Stereo, only without the noise reduction, and is found on consumer formats.

The principle of Dolby Stereo/Dolby Surround is that four channels, namely left, center, right, and surround are matrix-encoded down to the two channels that are subsequently called L_t (Left total) and R_t (Right total). These two channels are recorded on the two sound tracks of the films. When they are played back through the cinema's processor or the home system's Pro Logic decoder, the four channels are re-established.

A surround-encoded signal can also be listened to in normal stereo and in mono; however, without the benefits of the larger sound image. With stereo reproduction, all the information is present, only it is reproduced in the stereo system's two channels. In mono, the part of the signal that would have been reproduced in the surround channel disappears.

Dolby Surround has been made into a production format, not just for radio and TV sound, but also for CD releases, and in particular for sound in computer games. Even if you are not producing in the format, it is important to know how an arbitrary stereo production will be played back in the system since this often occurs.

Encoding

The simplest part of the process is the encoding. The principles the encoder operates under are as follows:

- Left is run directly to L_t.
- Right is run directly to R_t.
- Center is attenuated by 3 dB and then added in phase equally to L_t and R_t.

The surround channel is attenuated by 3 dB, band-pass filtered ($f_1 = 100$ Hz, $f_u = 7$ kHz), and low-level compression is added (modified Dolby B), passed through a 90° phase-shifting circuit, and run to L_t and R_t respectively in anti-phase.

The purpose of the 90° circuit is to make it possible to pan between center and surround. If it is not done in this manner, then the sound will go through half of the pan and end in either the left or right channel.

Mix & Mic

During the mixing process, attempts are normally made to use signals in the four channels that are uncorrelated in order to obtain the best possible channel separation in the subsequent decoding.

The stereo microphone technique that comes closest to producing a signal that corresponds to a matrix-encoded signal is the MS technique. There is a good correlation between the placement of the sound sources at the recording

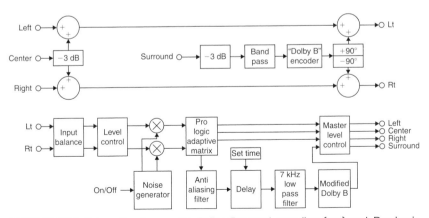

FIGURE 20.1 Schematic diagram for Dolby Surround encoding (top) and Pro Logic decoding (bottom). The simplest part is the encoding of left, center, and right since this can be done on any stereo mixer. The surround channel however must be inserted at ±90° due to a requirement of direct panning between center and surround.

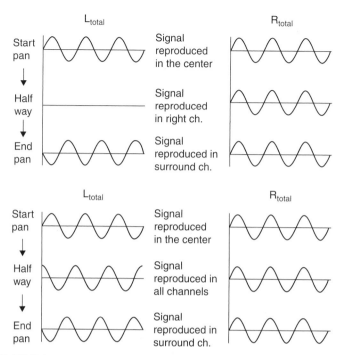

FIGURE 20.2 Upper: Panning from center to surround <u>without</u> use of 90° phase shifting. Halfway through the panning, the signal will lie in one of the side channels, here the right channel. Lower: Panning from center to surround <u>with</u> use of 90° phase shifting. Halfway through the panning, the signal lies in all channels. This is heard as the signal gradually disappearing out of the center and appearing in surround.

and the playback in matrix-encoded systems such as Dolby Surround, Circle Surround, and Lexicon Surround.

Decoding, Passive

In the simplest form of passive decoding that is found on certain sound cards, the four channels are created in the following manner:

$$
\begin{aligned}
\text{Left} &= L_t \\
\text{Right} &= R_t \\
\text{Center} &= (L_t + R_t) - 3 \text{ dB} \\
\text{Surround} &= (L_t - R_t) - 3 \text{ dB}
\end{aligned}
$$

This decoding is very imprecise due to poor channel separation, but it can, however, create a certain spatial effect.

Decoding, Pro Logic

With Pro Logic decoding, a procedure is undertaken based on the mutual level and phase relationships of the L_t and R_t signals, which causes signals that were solely represented in a single channel prior to the encoding to also appear only in that single channel after the decoding. For example, all center information will be removed from the left and right channels, all left information will be removed from center and surround, etc.

Before the surround information is sent out, the signal will be delayed (adjustable 20–100 ms) to take into account the Haas effect, thus avoiding localization to the rear speakers, which are typically placed closest to the listener. Then low-pass filtering is performed at 7 kHz in order for small phase errors in the transmission to not result in audible crosstalk in the channel. Finally, a noise reduction is performed in a circuit that corresponds to partial Dolby B with half effect (half expansion). The circuit affects signals below approximately -20 dB.

The Pro Logic decoder is fitted with a pink noise generator for adjusting the channel balance. Adjustments are made to the same level for each channel, measured at the listening position. All (both) surround speakers are counted as one channel.

The Pro Logic II decoder has been developed with the particular purpose of creating surround sound from normal stereo productions.

Important Details

The following details are important for the production and use of Dolby Surround to be able to judge levels correctly:

- It is important that all program material that may be decoded through a Pro Logic decoder is auditioned through one. Many stereo effects rely on relative phase effects and will thus end up in the surround channel.
- A 100 Hz bass cutoff occurs in the surround channel only in the encoder, not in the decoder.
- If L_t and R_t are mutually phase-shifted by 90°, but are otherwise at the same level, then the signal will distribute itself equally in all four channels.
- When mixing surround programs, the four channels must always be monitored via an encoder and decoder so that the spatial steering effects of the latter may be recognized and compensated for.
- Surround decoders for home use (Pro Logic) use a special terminology: *Normal*: Bass under approximately 100 Hz is filtered out from the center channel and distributed in the right/left channel (so one can use a center speaker with limited bass capabilities). *Wide*: All three front channels have full frequency range.
- In many computer-based games, Dolby Surround is used since the format is supported by a number of sound cards. Typically, a setup without a center speaker is used. In consumer decoders, this corresponds to *phantom center*, where the entire center signal is equally distributed to the left and right channels.

Loudspeaker Positioning — Dolby® Surround

The basic loudspeaker setup for Dolby Surround has three front speakers — left, center, and right. Normally, the center speaker is placed under the picture monitor, and if projection is used the speaker is placed behind the screen.

The surround channel is normally reproduced over two speakers placed diagonally behind the listener. In order to avoid the perception that the sound from the rear speakers is coming from a point between the two speakers, the signal to one of the speakers may be either phase-shifted or phase-inverted. When mixing in a small room, phase-inverting one of the surround speakers is not recommended as it will then be very difficult to judge the level in this channel. However, a 90° phase shift will work. In THX-specified systems for home use, it is possible to use dipole speakers to attain diffuse sound.

Acoustic Calibration

The decoder's pink noise generator is used in acoustic calibration. The electrical signal lies at −6 dB in relation to full modulation. Each channel is measured on its own. All speakers in the surround chain are regarded as one channel. In the listening position, a sound pressure level of 82 dB(C) is measured from each channel.

In a cinema (Dolby Stereo), the measurement is made over a larger area; typically four to five characteristic measurement points are selected. The

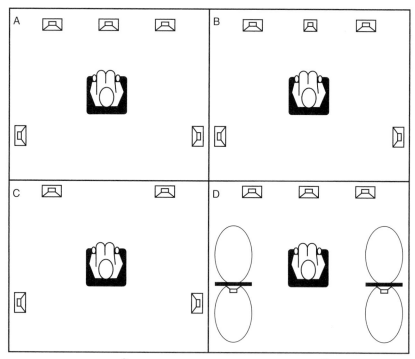

FIGURE 20.3 A: Dolby® Surround. Center mode: Wide. B: Dolby® Surround. Center mode: Normal (reduced frequency range in the center channel). C: Dolby® Surround. Center mode: Phantom (center is split between left and right channel). D: Dolby® Surround. The surround speakers are dipole speakers.

sound from each channel is then measured at these points and the result is an average of the measurements. Instead of a test generator, a special test film is used with pink noise recorded at −6 dB.

Pro Logic Decoders

Dolby Laboratories has granted a Pro Logic license to a number of manufacturers who produce processors. All Pro Logic circuits are approved by Dolby. Regardless, there are significant differences in the sound quality depending on which supplier has produced the processor concerned. The professional decoders and a more musical consumer version, Pro Logic II, are all produced by Dolby Labs itself.

CIRCLE SURROUND®

Circle Surround is another of the matrix-encoded systems, and is a 5:2:5 system. As opposed to Dolby Surround, this involves a system that splits the surround channel up into left and right surround. The format can also reproduce Dolby Surround encoded material.

LEXICON LOGIC7™

Lexicon has developed a decoder that is capable of extracting seven channels out of two. The system is primarily developed for music and has been introduced for car audio systems with three front channels, left and right surround/side and left and right back.

Digitally-Based Multi-channel Systems

While the matrix-encoded formats mainly build on analog technology, there is another group of more or less true multi-channel systems that are based on digital technology. What they have in common is that the channels are kept separate during the entire production phase until they are presented to the user. Here, they are packed together − sometimes with the use of bit reduction − into a single bit stream. The systems are used both for film sound in cinemas and for film and music production for listening in the home.

THE 5.1 FORMAT

Even though several other formats exist, most multi-channel sound production formats are based on the standardized 5.1 loudspeaker arrangement. The number refers to the fact that there are five channels with a full 20 Hz − 20 kHz bandwidth. In addition, there is one channel for sound effects at low frequencies. This channel is normally called the LFE channel (Low Frequency Enhancement or Low Frequency Effects). It has only a limited frequency range. Tomlinson Holman hence called it ".1" (even though the frequency range is less than 1/10 in comparison with the other channels). This name has stuck and it has subsequently been included in standards and similar official writings.

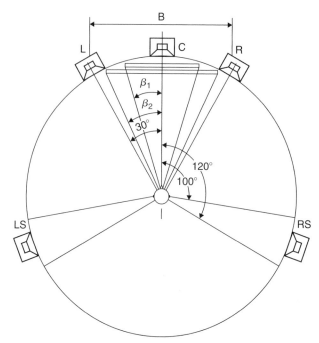

FIGURE 20.4 Listening setup according to ITU-R BS.775-1.

It does not involve an actual subwoofer channel or the like, because all primary channels go all the way down to 20 Hz. However, this LFE channel is amplified 10 dB more than the other channels to provide additional headroom. It may only be used for a bang that occurs once, half an hour into the program material.

The LFE or ".1" channel is included due to regard for the better utilization of the dynamic range of the primary channels. In general, it is used mostly in movies and only extremely rarely in music.

This configuration is also called 3/2 because there are three front channels and two rear channels. If the front speakers are in a row, it is suggested by the ITU that the center speaker is given a time delay. The LFE (or .1) channel has not been taken into account in this arrangement. The subwoofer can in principle be placed anywhere with appropriate regard paid to distance and acoustic conditions in the room.

Dolby® Stereo SR•D/Dolby® Digital

Dolby Stereo SR•D is a sound format on 35 mm film. It contains both analog and digital sound. The analog sound is encoded in Dolby Stereo with Dolby SR (Spectral Recording) noise reduction. The digital sound is encoded in Dolby Digital 5.1. The bit reduction system is Dolby AC-3 with a bit rate of 320 kbps. The LFE channel encompasses the frequency range 20 Hz − 120 Hz.

During playback in a cinema, a change can be made from the digital to the analog tracks if errors occur in the scanning of the digital information. On DVDs, Dolby Digital is used as one of the standardized formats, both on DVD-Video and DVD-Audio.

Surround EX

In connection with the recording of "Star Wars — Episode 1" Dolby and THX/Lucasfilm felt a need for an extra rear channel that could be introduced without any large technology-related problems. The center-surround channel is encoded into left surround and right surround using the same method used for the encoding of the center channel in Dolby Surround. The digital encoding and decoding remain unchanged in relation to Dolby Digital.

DTS® (Digital Theater System)

This format was, as the name suggests, developed by Digital Theater System, an American company. The bit reduction system used, Apt-X, was however developed in Ireland. The reduction has a fixed ratio of 4:1 (i.e., only 25% of the original quantity of bits remains). It is fundamentally a 5.1 format used both for film sound and music production. In both cases, the digital information is placed on a CD-ROM. The consumer version is called Digital Surround.

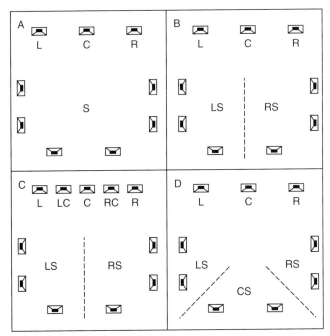

FIGURE 20.5 A: Dolby® Stereo (analog and matrix-encoded format). Configuration: 3/1 (three front and one rear channel). B: Dolby® SR•D (Dolby® Digital) and DTS (Digital Theater Systems). Configuration: 3/2. C: SDDS® (Sony Dynamic Digital System). Configuration: 5/2. D: Surround EX™ and DTS ES. Configuration: 3/3.

Arrangements as they occur in a cinema are as follows:

For normal feature films, DTS uses two CDs, which makes for a total feature length of 3 hours and 20 minutes. A timecode is printed alongside the analog sound tracks on the film. This timecode is used for synchronizing the CD-ROM player and projection machinery. The LFE channel here encompasses the 20 Hz − 120 Hz frequency range. The first film with DTS was Jurassic Park.

DTS-ES (Extended Surround)

DTS found it necessary to follow suit when Dolby developed Surround EX. Hence DTS is also able to offer a format with a center surround channel.

SDDS® (Sony Dynamic Digital Sound)

SDDS has eight channels: left, right, center, left center, right center, left surround, and right surround, and an LFE channel. It is a 7.1 format. SDDS differentiates itself from 5.1 by the fact that it uses two additional speakers placed between left and center and center and right. The system can, however, also function in 5.1 or 4/1, or 3/2 formats.

The bit reduction system is ATRAC (Sony's own), which is also used on MiniDisc. The compression is approximately 5:1. The maximum bit rate is 1411 kbps. SDDS is used only for films. The digital information is placed on the film itself.

DIGITAL CINEMA

The goal for the Digital Cinema is the establishment of completely file based formats to be distributed by data networks. Many cinemas have been refurbished to accomplish this goal. However, it has also been an investment to follow the trend. Actually in many countries the picture may have been improved by the transition to file based films. However, the play out sound systems have unfortunately gone from 5.1 to 4:2:4 matrix formats.

DISPLAY OF MATRIX-ENCODED SIGNALS

The two channels containing matrix-encoded signals can be monitored on normal stereo instruments. The goniometer is particularly useful because it can show whether the phase angle is too wide, etc. The goniometer is incredibly efficient when it comes to the quick overview. It is also important to monitor the balance between the L_t and R_t channels. Even though most decoders are relatively effective in maintaining a center impression, imbalance may result in an incorrect division of the channels. It is also possible to use an instrument with a built-in surround decoder.

VIRTUAL SURROUND

Virtual surround is a set of techniques where only two speakers or a set of headphones are used to recreate a sound field from many speakers. The effect is

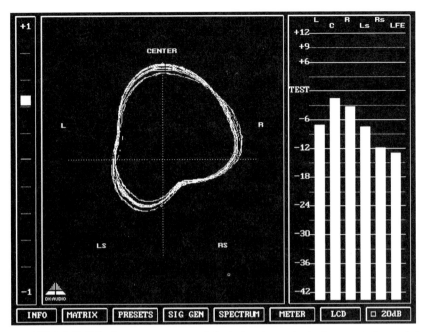

FIGURE 20.6 This shows one manner of displaying the content of a surround-encoded signal. The signal is strongest in the "most full-bodied" direction. This display is popularly called a "Jelly Fish™." At the same time, the bar graph meter shows the precise channel level. This instrument, the MSD600, can display both Pro Logic and true multi-channel audio.

often arbitrary. The surest way to monitor signals is to use the goniometer since the signals typically have high phase opposition content.

DISPLAY OF 5.1

With true 5.1 or the equivalent, the channels will be kept separate prior to the final coding into one of the standardized formats. Six bar graph meters can of course show what the individual channels contain. However, it is most practical and more manageable to use an instrument with a Jelly Fish™-like display.

Jelly Fish

Jelly Fish™ from DK Technologies is the popular — and registered — name for a goniometer-like display on a screen.

The figure in itself does not show the phase between the channels, but rather the amplitude in each of the channels concerned. The purpose of this display is to create an overview.

If the incoming signal is matrix-encoded, a decoding is performed first of L_t and R_t so that the resultant channels are obtained. If it is true multi-channel, then the signals are used directly. On the instrument's screen, a circle is fundamentally established and the magnitudes of the levels in the channels concerned (left, center, etc.) are multiplied into this figure. The figure thus becomes the "most full-bodied" in the direction/channel that has the strongest signal. For the sake of

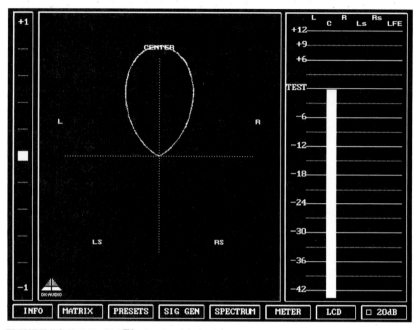

FIGURE 20.7 Jelly Fish™: signal solely in the center channel.

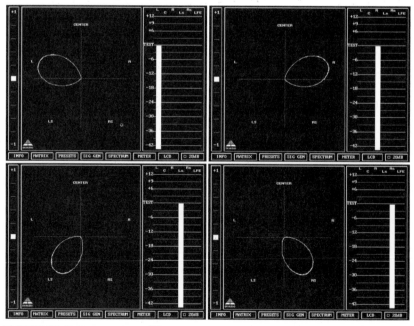

FIGURE 20.8 Upper:Jelly Fish™: signal solely in the left and right front channels. Lower: Jelly Fish™: signal solely in the left and right surround channels.

clarity, the instrument will normally have a certain inertia in order for the user to be able to follow that part of the signal that has a certain weight in terms of time.

Phase Differences between Adjacent Channels

Since the figure does not show anything about the phase, Jelly Fish™ provides a change of color in the transition area between two channels if the phase angle is greater than 90°.

ACOUSTIC CALIBRATION OF MULTI-CHANNEL SYSTEMS

Calibration of the acoustic sound levels has been a requirement for many years when working with sound for film, although it has not been particularly common in other branches of the sound industry. However, with the widespread use of multi-channel formats for all forms of music and film presentation in the home, it has turned out to be beneficial to also calibrate the acoustic levels for these formats.

It is important to differentiate between production for the cinema and production for 5.1 channel reproduction in the home based on ITU 775.

CALIBRATION OF CINEMA SYSTEMS

In a cinema, the listeners sit far from the speakers. Presumably, the majority sit in the diffuse sound field. In any event, attempts are made to establish a diffuse sound field from the surround speakers. Thus, when the sound pressure is measured inside the cinema or in a mixing theater it must be averaged over many different measurement locations. The typical basis for the majority of standards is at least four locations. If there are different areas for the audience, for example main floor and balcony, measurement should be made in at least four locations in each.

Before performing this measurement, the system's frequency response must be in order. Normally, the ISO 2969 X curve standard is used as a measure for the characteristics of the system.

Optical Sound

Cinema systems for the reproduction of Dolby Stereo (analog optical sound), normally have a built-in generator with pink noise. This signal is sent out at a level corresponding to half modulation of the optical track (i.e., 6 dB below full modulation).

The generator is used in particular in the mixing theater, where the sound has of course not hit the recording media yet. With this, the B chain can also be checked; that is, that portion of the sound system that encompasses everything from the playback system for the specific cinema up to and including the acoustic space. (The A chain encompasses that portion of the system that lies before the playback system for the specific cinema.)

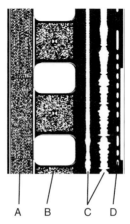

A   B   C   D

FIGURE 20.9 A: SDDS (corresponding tracks are found on the opposite edge of the film). B: Dolby Digital. C: Analog sound tracks. D: DTS sound tracks on 35 mm film.

A test film (Dolby cat. No. 69) with prerecorded pink noise at 6 dB under full modulation and with Dolby noise reduction is run in the cinema's projector in order to assist in making adjustments to the B chain. For each of the four channels (L, C, R, and S) adjustments are made for a sound pressure level of 85 dB(C) (integration time: slow) in the inside of the cinema as calculated by a simple average value of the measurement results at the selected measurement locations. This procedure regards a chain of surround speakers as one channel (i.e., all the speakers in this chain must be operating at the same time).

Digital Sound

Digital sound on film has created a larger dynamic range, of which a large part is used for greater headroom in comparison with optical sound.

The digitally recorded SMPTE standardized test signal (pink noise) lies at −18 dBFS. During playback of each of the front channels, this signal must be reproduced at a sound pressure level of 85 dB(C). The two surround channels are each adjusted to 82 dB(C). This causes the level created by the entire surround chain to thus equal 85 dB(C).

A 10 dB amplification is inserted in the playback chain for the LFE signal. When the limited bandwidth (20 Hz − 120 Hz) pink noise from the LFE channel is played back, it is possible with a 1/3-octave spectrum analyzer to see that the individual ranges in the LFE channel are reproduced 10 dB higher than the individual ranges in each main channel. Measured as a C-weighted sound pressure level, the LFE channel will show a level that is approximately 4 dB(C) higher than the level in each main channel.

CALIBRATION OF 5.1 IN AN ITU-775 ARRANGEMENT

In a 5.1 system based on the ITU arrangement, all main channels have in principle the same conditions: there is one speaker per channel and each is placed the same distance from the listener.

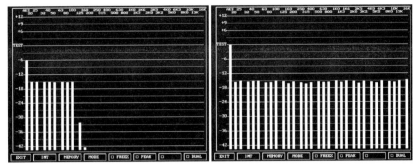

FIGURE 20.10 The spectrums are shown here for pink noise recorded on LFE (on the left) and on a main channel (on the right). The LFE only goes to 120 Hz, whereas the main channel has full bandwidth. All columns in the active areas of the channels are of equal height (i.e., the channels have the same level per 1/3-octave). The column furthest to the left in each spectrum shows the total level. The level is 6 dB lower in the LFE channel than in the main channel. This is because there is a smaller frequency range represented here.

Internationally, there is however agreement neither on the level nor on the bandwidth for the noise signal that is used for acoustic calibration. Pink noise is good since it includes all frequencies; however, it is impractical due to its "unsettled" character, which makes it difficult to measure at low frequencies.

SURROUND SOUND FORUM

Surround Sound Forum (SSF) is a German interest group established by Verband Deutscher Tonmeister (VDT, Association of German Tonmeisters), the Institut für Rundfunk Technik (IRT, Institute of Broadcast Engineering), and Schule für Rundfunktechnik (SRT, School of Broadcast Engineering). The SSF has prepared guidelines that are generally accepted in Europe. Three test signals are specified in it, which are recorded at −18 dBFS (RMS).

TABLE 20.1 Measurement signals for the main channels in surround sound configuration

Signal(only in one channel)	Measurement signals			Listening level	
	PPM level with $\tau<0{,}1$ ms[dB]	PPM level with $\tau<10$ ms[dB]	RMS level[dB]	Sound pressure level SLOW [dB]	Sound pressure level SLOW [dB (A)]
1 kHz sine	−18	−18	−18		
Pink noise 20 Hz − 20 kHz	−9	−13	−18	82	78
Pink noise 200 Hz − 20 kHz	−11	−15	−20	80	78

SMPTE

The corresponding standard from SMPTE (RP155) uses a standard of -20 dBFS for the reference level. Here, the C-weighted sound pressure level ends up at 83 dB.

BASS MANAGEMENT

Bass management consists primarily of filtering out the bass from the main channels and reproducing it in a subwoofer (together with the LFE signal). Frequency response and level must display the same data as if only full range systems were being used in the main channels.

OTHER SYSTEMS FOR SURROUND SOUND

New systems are being developed all the time. Some of these systems are intended for commercial use in the home and other systems are more likely to be used in a fixed installation in venues for theater, special events, etc.

TMH 10.2 is a system developed and demonstrated by Tomlinson Holman. It involves speakers over the head.

NHK 22.2 is a Japanese system developed by the national broadcaster with 24 audio channels intended for broadcast. It has been demonstrated and programs have been made for this format.

Sound Field Synthesis is being developed and promoted mainly in the Netherlands by Diemer de Vries and his team at the Delft University. The system recreates the soundfield. One system can easily consist of 64 channels.

UPMIXING SYSTEMS

Most software based mastering tools include algorithms to upmix from stereo to surround, both 5.1 and 7.1. This has been a necessity in order to provide content for the many systems found on the market. These algorithms are not all alike and must be assessed in connection with the program material to be converted.

Chapter | twenty-one

Standards and Practices

CHAPTER OUTLINE

Audio Metering. DOI: 10.1016/B978-0-240-81467-4.10021-8
Copyright © 2011 Eddy B. Brixen. Published by Elsevier Inc. All rights reserved.

151

When program material is exchanged, it is of course practical that both the sender and the receiver know how the levels and channels are placed. It is easiest if there is a standard that the largest possible number of people agree to follow. Sometimes the standards are set internationally. Other times they are just based on common practices. Normally it is possible to see a difference between practices in Europe and in the US.

A listing of some of the applicable standards and practices follows.

ANALOG TAPE, AUDIO

Even though the majority of recordings today are stored in digital formats, the analog tape recorder is not completely dead. If nothing else, there is some very good archive material stored in the vaults. It would thus be nice to know how to play it back.

1/4 in., 2 Track

STEREO, LEFT/RIGHT
Track 1	Left
Track 2	Right

STEREO, M/S
Track 1	M(iddle)
Track 2	S(ide)

1/4 in. or CC, 4 Track

SLIDE-SHOW
Track 1	Left
Track 2	Right
Track 3	Not used
Track 4	Time code/cue

Perforated Magnetic Film, 4 Track (DIN 15.554 / ISO162)

DOLBY STEREO
Track 1	Left
Track 2	Center
Track 3	Right
Track 4	Surround

Perforated Magnetic Film, 6 Track (Todd AO)

Track 1	Left
Track 2	Center

Track 3   Right
Track 4   Surround
Track 5   Outer left
Track 6   Outer right

Multi-track Tape

No actual norms exist for the placement of material on specific tracks on multi-track machines, since we are not dealing with master tapes. However, there are a couple of practical rules that should be adhered to for technical reasons.

With sync-recording: Never record on a track adjacent to the track that is being listened to.

With the recording of stereo tracks: Stereo tracks are always placed beside each other, for example tracks 1, 2 or 17, 18. In connection with audio-audio synchronization: Never place two stereo tracks on separate machines.

Time code: Always placed on the outside track, as a rule the track with the highest number.

Test Tones

The test tones, when properly used, are the key to painless editing and subsequent correct playback. On the finished program, the tones represent what can be expected during the rest of the program.

On analog audio tape, including perforated magnetic film, the tones always refer to a level of magnetization, which is typically specified in nanoweber per meter of track width, abbreviated as nWb/m. If not otherwise specified, then 1 kHz sinusoidal tones are used. This level is also called the nominal level. In addition, there will be headroom, which is determined by the level at which the distortion on the tape exceeds a certain magnitude, for example 3%.

35 mm Magnetic Film (perf)

320 nWb/m (Europe)
185 nWb/m (USA)

2 in., 24 Track

(76 cm/s or 30 in/s)
510 nWb/m (Europe)
355 nWb/m (USA)

1/4 in., 2 Track with Dolby SR

320 nWb/m (Europe)
200 nWb/m (USA)

1/4 in., 2 Track

(38 cm/s or 15 in/s)
510 nWb/m
320 nWb/m (Europe)
200 nWb/m (USA)

1/4 in. Full Track, Nagra

(19 cm/s or 7½ in/s)
320 nWb/m (Europe)
200 nWb/m (USA)

Compact Cassette

(@ 315 Hz)
250 nWb/m (Europe)
200 nWb/m (USA)

ANALOG TAPE, VIDEO

Professionally, it is mainly BetaCam that is relevant here. The one-inch formats are predominantly used for archiving IF the programs have still not been digitized.

Video, 2 Track

Mono program, master: all formats except U-matic
Track 1   Final mix

U-matic
Track 2   Final mix

Mono, for voice-over: all formats except U-matic
Track 1   Final mix
Track 2   (IT: International Track)

U-matic
Track 2   Final mix
Track 1   IT

Comments

"IT" or "I sound" can contain M (music), E (effects), or D (dialog, in-vision), but never a feed line (dialog where the speaker is not in the picture). Hence, a track can also be marked for example as M+E, M+D or M+E+D.

Stereo, left/right
Track 1   Left
Track 2   Right

Stereo, M/S

    Track 1   M(id)
    Track 2   S(ide)

Test Tones

On analog video tapes, the test tones will refer to a magnetization level on the linear, longitudinal track, whereas a frequency fluctuation ought to be referred to either in kHz or in percentage for the AFM tracks.

3/4 in. Umatic
100 nWb/m

Betacam SP, Linear
100 nWb/m

Betacam SP, AFM
−9 dB re 100% mod.

SVHS, Linear
100 Wb/m

SVHS, HiFi (AFM)
−9dB re 100% mod.

Broadcast, Programs

Test tone	1 kHz
Level	6 dB below full modulation. On a PPM instrument with the Nordic scale the modulation should be 0 dBu
Duration	90 seconds
Start, time code	00:00:00:00 (or 09:58:00:00)
Start, program	00:02:00:00 (or 10:00:00:00)

Stereo

Left/right: Test tone first alone in left channel for approximately 5 seconds, then both tracks for 90 seconds.

Note

The EBU test tone interrupts the left channel on a regular, repeating basis. The BBC employ a bespoke stereo identity tone (GLITS) that interrupts the left channel once, followed by two interruptions on the right channel − the three interruptions being equally spaced and repeated on a regular basis.

M/S: 1 kHz, 6 dB below full modulation in both channels. The channels are adjusted before matrix-encoding for left and right

Note

If the test tone is recorded as left and right, and thereafter matrix-encoded to M $((L+R)/\sqrt{2})$ and S $((L-R)/\sqrt{2})$ on the tape, then there will only be a signal on the M track, which will be 3 dB above the test level. The S track should then be completely without a signal.

Note:

For analog formats, the test tone ought to always be recorded without noise reduction if it influences the recorded level. It should be noted on the tape report whether the test tone was recorded with or without noise reduction (for example, "-NR").

Noise Reduction

The different noise reduction systems for analog tape formats each have their own special adjustment procedures, which of course must be adhered to in order to attain the optimum dynamic range. In the following, we will discuss some of the test signals used in existing formats.

Dolby A

Tone:	850 Hz FM modulated with ±10% per 0.75 seconds
Duration:	30 seconds
Level:	Normally modulated to a magnetization level of 185 nWb/m regardless of the format if nothing else is specified. However, this is highly dependent on the type of magnetic tape used. On standard audio tape the level is set to 320 nWb/m. For the 1 inch video formats the level was 100 nWb/m.

Dolby B

Tone:	400 Hz sinusoidal, FM modulated with ±10% per 0.75 seconds
Duration:	30 seconds
Level:	Full modulation

Dolby C

Tone:	400 Hz sinusoidal, FM modulated with ±10% per 0.75 seconds
Duration:	30 seconds
Level:	Full modulation

Dolby SR

Tone:	Pink noise, interrupted for 20 ms every 2 seconds
Duration:	30 seconds
Level:	15 dB below full modulation

Telecom c4

Tone:	550/650 Hz, alternating every half second
Duration:	30 seconds
Level:	Full modulation, or 4 dB below full modulation

DIGITAL TAPES, AUDIO

The determining factor that decides the level on digital tapes is based on the use of the program material. Typically, there will be a difference between a broad-casting master at a broadcast level (allowing typically 9 dB headroom) and a master for a CD or DVD (with zero headroom).

EBU

When exchanging programs between radio broadcasters, it is important that the tapes can be played directly. Hence under the auspices of European cooperation within the EBU, the following standard has been adopted:

Test level: -18 dBFS
Max level: -9 dBFS

Hence, there is so-to-speak 1½ bits (9 dB) that are not used. This was primarily done for reasons of safety, in order to avoid any risk of overloading, in particular when the signal levels may have been monitored with analog metering systems.

In the Loudness recommendation (see later) -1 dBTP is allowed (if measured using oversampling).

CD Master

Old vices from the vinyl era remain true to form: the level is set as high as possible. In pop/rock productions, a liberal view is taken of digital overloading. Devices are actually manufactured that deceive the mastering system into believing that no overloading is occurring by moving all encoding levels on full scale one step down.

Apart from that, all samples at 0 dBFS are normally logged. Or to be more precise: if a number of samples in a row are at full level, then the time code is noted before they are delivered for mastering.

Post-production

Tapes to and from post-production ought to adhere to the shown allocation of channels. It is current in both Europe and the USA.

8-track digital media (for example DA88)

Track	1	2	3	4	5	6	7	8
Channel	L	R	C	LFE	LS	RS	20 bit data	20 bit data

DVD/Blu-ray/DTV master

Track	1	2	3	4	5	6	7	8
Channel	L	R	C	LFE	LS	RS	LT	RT

Test signal: 1 kHz sinusoidal
Level: Europe: -18 dBFS
            USA: -20 dBFS

DIGITAL TAPES, VIDEO

Tracks and equipment follow digital audio in the respective areas (Europe/USA).

BLITS

BLITS (Black & Lane's Identification Tones for Surround) provides channel identification tones for all channels within a 5.1 Surround Sound signal. This also helps to provide information in a stereo downmix — channel presence/absence. The BLITS tones can be used at the start of a program to help identify channels, in OB trucks for channel identification back to the studio and MCR as well as lineup for storage devices.

The frequencies used are based on the international music standard and are L = 880 Hz, R = 880 Hz, C = 1,320 Hz, LFE = 82.5 Hz, Ls = 660 Hz & Rs = 660 Hz. The arrangement of the tones is designed to provide sequential and easy to read displays on bargraph meters. Basically, this sequence can be generated by the metering system if the software is available.

HARD DISK SYSTEMS / FILE BASED SYSTEMS

There is no agreement among the producers of hard drive systems on how analog and digital levels should go together. Test levels are placed at −18, −19, and −20 dBFS.

In a working situation, −6 dBFS (corresponding to approximately 50% of full amplitude) should normally not be exceeded. However, new metering systems offering reading of true peak (using oversampling) extend this to −1 dBTP for linear PCM. If, however, the audio is bit reduced the headroom must be enlarged as some systems may produce higher peak levels when converted.

Hard disk based audio work stations often have a useful function called "normalization." The signal is adjusted to a specific level, defined either by dB re full scale or by a percentage of full modulation. It should be noted that the reference is nearly always made to the amplitude of the sample that has the highest level and not to the RMS level or, for example, loudness.

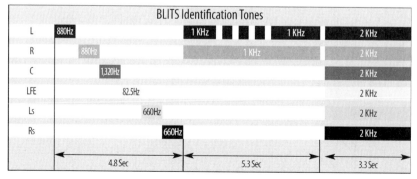

FIGURE 21.1 The BLITS — Identification Tones for Surround.

In file based systems the Meta Data is very important. This leaves a possibility of batch normalization of the files without actually changing the files themselves. What is changed is a number in the metadata. Thus the play out systems can correct to standardized signal level or loudness level.

MASTERS FOR VINYL

There are some (good old) rules that must be adhered to in order to make a master for subsequently cutting records (particularly if the master is produced from a digital medium).

1 Avoid excessive bass, especially below 80 Hz.
2 Avoid excessive treble.
3 Avoid excessive stereo separation.
4 Avoid too large a dynamic range.

These rules of thumb are due to the limitations of the cutting technology. One should know, for example, that the recording speed is different at the outer and inner peripheries. This is of significance to both the dynamic and the frequency response.

Analog Masters

30 s.	Leader tape, white + red
10 cm [4 in.]	Leader tape, transparent
10 cm [4 in.]	Leader tape, white + red
5 s.	1 kHz, left channel. 0 VU: 320 nWb/m
30 s.	1 kHz, both channels. 0 VU: 320 nWb/m
30 s.	10 kHz, both channels, −10 dB re 0 VU
30 s.	100 Hz, both channels, 0 VU: 320 nWb/m
(30 s.	NR ref-signal (if NR is used)
10 s.	Leader tape, white + red
	First recording
4 s.	Leader tape, blue
	Second recording
4 s.	Leader tape, blue
	etc.
	Last recording
5 s.	Leader tape, red
10 cm [4 in.]	Leader tape, transparent
30 s.	Leader tape, red

Digital Masters

Digital masters follow the same principles as the analog masters (without leader tape of course). However, one needs to be really attentive to the general rules for good mastering for vinyl. The limitations of the record are not known in the same way in digital technology.

FILM, OPTICAL

The sound camera has a clipping level that is determined by the physical limits of the optical track. The test level is placed at 50% modulation, 6 dB below full modulation. This level also corresponds to the dialog level.

FILM, DIGITAL

Digital films will follow the SMPTE standard, that is, dialog (test level) will be at −18 dBFS. This corresponds to a sound pressure level of 85 dB(C) for each of the front channels.

FROM FILM TO DVD OR BLU-RAY

The acoustic sound level from test and dialog levels of digital film sound channels lies at −18 dBFS. This is adjusted in the cinema to a sound pressure level of 82 dB(C), each channel by itself, so that the total level from the two surround channels becomes 85 dB(C). In a home cinema, the adjustment process is such that all channels are adjusted to the same listening level. This means that the surround here will be too strong in relation to the balance in the cinema. Hence, it is normal that each of the surround channels is attenuated by 3 dB before they are transferred to DVD, Blu-ray, or digital TV.

The standard listening level for −18 dBFS (pink noise) is normally set to a sound pressure level of 78 dB(C) for each of the five main channels. When all five channels are playing at the same time, this then corresponds to a sound pressure level of 85 dB(C).

SATELLITE

For an uplink to a satellite, there can be different standards depending on the system in use. What is most important for analog connections is to check whether pre-emphasis is included in the instrumentation's display. If not, there is a risk of bringing the connection down, even if there is a program limiter on the output.

BROADCAST, TRANSMISSION

The ITU has established a test signal for transmission lines. This test signal contains three levels, which are defined here and related to EBU R68.

Measurement level (ML) is the level at which measurements are performed on the lines. The level corresponds to −30 dBFS.

Alignment level (AL) corresponds to the test level or nominal level. The level corresponds to −18 dBFS.

Permitted maximum level (PML) is the maximum that can be sent on the line. The level corresponds to −9 dBFS.

BROADCAST, PROGRAM

Recommendations for program levels are defined by the individual broadcaster. In pure analog circuits this will normally imply level setting in accordance with the EBU R68 as mentioned above. This may apply to digital transmission as well. However, it is expected that the loudness related recommendations gradually will take over. Broadcasters are re-writing their technical specifications for program delivery. Two sets of recommendation are in play here. Both have the ITU 1770 as the basis by which the loudness level is stated by one single number; nevertheless the program contains mono, stereo or surround sound.

ATSC Document A85:2009

Advanced Television Systems Committee Inc. has worked out a set of guidelines intended for the Dolby AC3 based audio. Originally the level settings were determined by a more complex measure of the Dialog Level, providing data for the dialnorm (dialog nomalization). These data are sent as metadata along with the program to provide the optimum setting at the receiver. However, after the introduction of the ITU 1770 on program loudness the k-weighting algorithm has been a valid method to provide data of Dialog Level for the dialnorm setting.

The target loudness is defined as -24 LKFS. The max peak is -2 dBTP (the max level of the true peak is -2 dBFS).

EBU Recommendation R 128

The EBU recommendation R 128 is based on the ITU 1770. However, a gating function has been added, providing a slightly different target for the loudness compared to the ATSC guidelines.

The program Loudness Level shall be normalized to a Target Level of -23 LUFS.

Permitted deviation is ± 1 LU. The Loudness Level shall be measured with an instrument according to ITU 1770 and EBU Tech Doc 3341.

The Loudness Range shall be measured with a meter compliant with EBU Tech Doc 3342 involving a gate level 8 LU below the Target Loudness. (The Loudness Range is program dependent.)

The Maximum Permitted True Peak Level of a program during production shall be -1 dBTP.

Please note that the assessed loudness is not only dependent on the program loudness. It also depends on how channels are routed in the receiver. If the program for instance has one single mono channel — and is normalized according to the loudness of this — and then is played back by the receiver in two channels, the perceived level is too loud.

INTERNET

The practice is that sound files for playback over the Internet must be compressed and normalized to full level before they are converted to the file type that is placed on the server. This type of file will normally be some sort of bit-reduced format for streaming at available bandwidth.

Bibliography

ATSC Recommended Practice: Techniques for Establishing and Maintauning Audio Loudness for Digital Television. Document A/85:2009.

EBU Recommendation R 128 Loudness normalisation and permitted maximum level of audio levels.

EBU Tech Doc 3205-E. 'The EBU standard peak-programme meter for the control of international transmissions'.

ITU-R BS.645. 'Test signals and metering to be used on international sound programme connections'.

ITU-R BS.1770. 'Algorithms to measure audio programme loudness and true-peak audio level'.

EBU Tech Doc 3341 'Loudness Metering'.

EBU Tech Doc 3342 'Loudness Range'.

EBU Tech Doc 3343 'Practical Guidelines for Production and Implementation'.

EBU Tech Doc 3344 'Practical Guidelines for Distribution'.

Summation of Audio Signals

CHAPTER OUTLINE

The summation of audio signals can be acoustic or electrical. Acoustically, it can be two sound sources in a room recorded at one position. Electrically, it can be two microphones that are connected to different inputs on a mixer. These signals are subsequently combined electrically and then routed to the same output.

The total level of the combined signal depends on the nature of the two signals. Some characteristic examples follow.

ELECTRICAL SUMMATION, CORRELATED SIGNALS

When signals are correlated, it means that they generally resemble each other or that they have much in common.

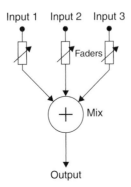

FIGURE 22.1 A sound mixer can be regarded as a summation component.

Audio Metering. DOI: 10.1016/B978-0-240-81467-4.10022-X

If we imagine that a test tone is recorded on a tape recorder or another medium, for example a 1 kHz sinusoidal tone in both the left and right channels, then the signals in the two channels are correlated. In fact, they are identical: the same phase and the same level. If these signals were combined electrically into one signal, the level would be twice the level of the individual channels. Specified in dB, a doubling of the level is the same as 6 dB. (See Chapter 6 on dB.)

If, however, you have two identical signals and one of them is phase-inverted (or the phase is turned 180°) before the summation, then the result will be 0 — in other words nothing. Expressed in dB, the result would be $-\infty$ dB.

ELECTRICAL SUMMATION, UNCORRELATED SIGNALS

If the signals were recorded on the two tracks of a tape recorder, then there would be some noise in the form of hiss. This noise is certainly of the same magnitude on both tracks; however, it is random and hence uncorrelated. By summing the two tracks the noise level will therefore only be increased by 3 dB.

Among recording engineers, it has always been a familiar trick to record two tracks if a stereo machine is available, even if only a mono recording is to be made. The signals add to +6 dB, but the noise only adds to +3 dB. In this manner, it is possible to improve the signal-to-noise ratio by 3 dB.

SUMMATION FOR MONO

It can be practical to use a mono converter if two channels of a stereo signal are to be combined. This is the still the case when TV programs with stereo sound are to be broadcast. In some parts of the world, NICAM is used for the stereo sound; however, the original FM carrier must also be provided with sound in mono.

The simple mono converter is built so that it only takes the levels into account, in other words:

1) For stereo, random distribution of the signals: mono is L+R − 3 dB.
2) For the same signal in both channels (or highly correlated signals): mono = L+R − 6 dB.
3) A signal in only one channel: mono = L or R.

However, this form of combination does not take into account program material, where there can be significant information stored in the opposite phase. This is the case with surround programs, which are matrix-encoded down to two channels. The same applies for stereo recordings that build upon MS. Or produced stereo where the width of the acoustic image is attained by a large degree of opposite phase content from effect processors and effect devices. In these situations, the use of 90° summation is relevant.

90° Summation

For 90° summation, one of the signals is shifted by 90° at all frequencies. This means that one signal is delayed in time with respect to the other one. However,

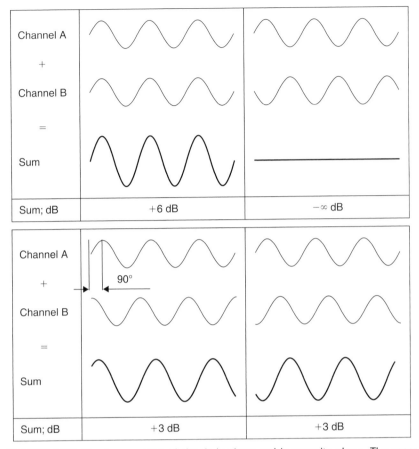

FIGURE 22.2 Direct summation of signals in phase and in opposite phase. The upper part of the illustration shows the direct summation. The lower part of the illustration shows the summation by the use of a 90° phase shift, where signals in both phase and opposite phase will add to the same level.

instead of a time delay that is constant at all frequencies, this involves a delay that depends on the frequency.

Time Delay in a 90° Phase-Shifting Path

At 20 Hz one of the signals is in principle delayed $1/20 \cdot (90/360)$ s $= 12.5$ ms.

At 20 kHz one of the signals is in principle delayed $1/20.000 \cdot (90/360)$ s $= 12.5$ µs.

The phase relationship between two signals **a** and **b** is significant to the final level after the summation when the following applies:

$$a = b \cdot \cos\varphi$$

where
$\varphi =$ the phase angle (the phase between the two signals)

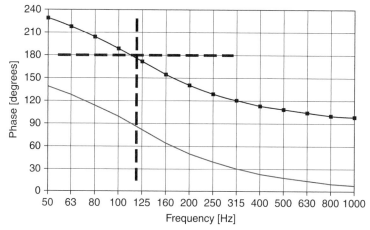

FIGURE 22.3 An example of one of the very few situations where a 90° summation has unfortunate consequences: Two signals are to be combined together, but there is an analog low-cut/HP filter on one of the channels. This is not clearly heard in stereo, but the channels are in opposite phase at around 115 Hz, which sounds like (and is) a lack of bass in mono.

COMB FILTERING

The filtering function that arises when a signal is added to itself after having been delayed in time is called a **comb filter**. The resulting frequency response resembles a comb, hence the name. The comb filter function is almost never intentional, but it is heard all the time in sound productions, where it can arise both acoustically and electrically.

Acoustically, it typically occurs when the sound on its way from source to recipient takes in part a direct path and in part an indirect path via a single reflective surface. The reflection must be attenuated at least 10 dB and preferably 15 dB in order for it not to have an effect on the sound field at the recipient position. Electrically, the phenomenon arises when two microphones with a certain distance between them capture the same signal and the level from each microphone is of the same order of magnitude.

In general: All digital signal processing takes time. This means in practice that comb filter effects can arise if you loop a signal via, for example, a compressor and combine this signal with the original.

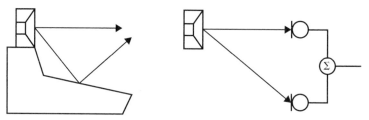

FIGURE 22.4 Two typical situations in which comb filters arise, either acoustically or electrically.

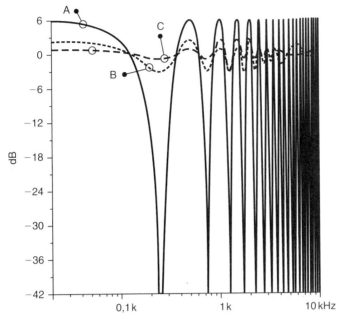

FIGURE 22.5 An example of a comb filter created by the combining of two signals with the same amplitude, but with a time delay between them of just 1 ms. A dip occurs due to cancellation at 500Hz, 1.5 kHz, 2.5 kHz, etc. The two signals add to double their value (+6 dB) at low frequencies and with a full wavelength's delay at 1 kHz, 2 kHz, 3 kHz etc.

Dip Frequencies

Cancellation occurs for a comb filter at all the frequencies where the two signals are in opposite phase. This occurs when the time delay comprises of periods of a duration of ½, 1½, 2½, etc. At 1 kHz the period is 1 ms; half of the period is 0.5 ms. If a time delay of precisely 0.5 ms occurs, it means that cancellation will arise, not just at 1 kHz, but also at 2 kHz, 3 kHz, 4 kHz, etc.

ACOUSTIC SUMMATION OF SIGNALS

The total sound level by the acoustic summation of two sound sources, for example two monitor speakers, depends on both the signal and the acoustics. The sound sources can be correlated or uncorrelated, as mentioned previously.

The listening position (or measurement position) can be either in the direct sound field or in the diffuse sound field. In the direct sound field, there is only one sound direction. This direct field exists either in the open, in a reflection-free room, or close to the speakers. The diffuse sound field occurs in a room when you are so far away from the speakers that the portion of direct sound is less than the sum of all the reflections. The distance from the speakers where the direct sound field and the diffuse sound field are equally large is called the **critical distance**. In a control room, it can typically be 1—3

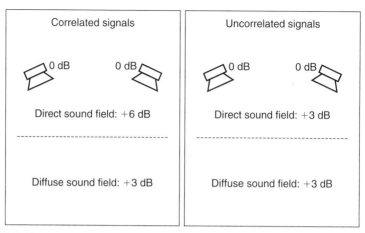

FIGURE 22.6 Summation of sound from two sources with the same level. The summation in the direct field will be determined by whether the sources are correlated or not. The summation in the diffuse field is independent of the correlation of the sources.

meters; however, it varies with frequency. At lower frequencies the critical distance is shorter compared to that at higher frequencies. The near field in front of the speakers can be regarded as a direct field.

CHAPTER OUTLINE

When digital information is to be transferred from one device to another, a protocol is used to ensure that both the transmitter and the receiver are in agreement on which bit means what. In addition, the physical connection is specified for connecting the two devices. In doing this, an interface has been defined.

PROTOCOL

Each sample, which is described by a number of bits arranged in a sequence, can have a number of extra bits (metadata) appended to it that give information associated with the sample concerned. It could be information about which channel the signal belongs to, whether pre-emphasis has been applied on the analog signal, data on the actual sampling frequency, or a lot of other useful information. Samples and extra information are combined into frames. The protocol defines in part the structure of the individual frames and the combining of frames into blocks of data.

Audio Metering. DOI: 10.1016/B978-0-240-81467-4.10023-1

PHYSICAL CONNECTION

The digital information is transferred in the form of a voltage signal, a current signal, or as light impulses on a fiber optic cable. The individual standards define what method or methods are used.

Even though it is sound that is being transferred, the frequency content in such a connection will immediately reach up to several MHz. Hence sound signals transferred as digital signals should always be treated as RF.

AES3

The most widely used interfaces for audio were originally initiated and standardized by the Audio Engineering Society and the European Broadcast Union, and hence used the names of both of these organizations. However, the AES has taken over the maintenance of this standard, so the name has just become "AES3." It is a standard that is constantly under development as new options in digital audio are brought to the market.

The format is serial (i.e., even though it contains two channels, it can be run on one wire). During the transmission, a sample from the first channel is transferred, followed by a sample from the second channel and so on.

The physical connection is comprised of a balanced 110 ohm cable with XLR connectors. The level is 2−7 Vpp (peak to peak). However, the signal can be read all the way down to 200 mVpp. Transport lines other than balanced lines may be applicable.

The core of the data portion for AES3-2 consists of 20 sound bits, and 4 extra bits (aux). These 4 extra bits have gradually come to always be used for sound, so regardless of how many bits are used in the sampling, 24 bits per sample are transferred. For example, if 16 significant bits are involved (as in the CD format) then bits 17-24 will simply be set to the value "0." Each sample (from either the right or the left channel) is placed into a 32-bit sub-frame.

PREAMBLE AND V, U, C, AND P

The eight "surplus" bits in each sub-frame are used for synchronization and extra information. The first four bits comprise a preamble, which indicates which channel the current frame represents or whether it is the first sub-

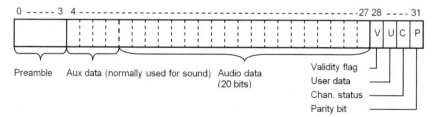

FIGURE 23.1 AES sub-frame, "packaging" of each sample. Each sub-frame begins with a preamble to help with the synchronization. Each sub-frame ends with four bits containing useful information.

frame in a block. The last four bits contain extra information that can say something about the signal that is being transmitted. This information is divided up into four individual bits, which are designated V, U, C, and P.

V stands for Validity. This bit indicates whether the associated sample is in order. P stands for Parity. This parity bit is set so that an even number of 1s is always attained in each sub-frame. If a check on the Parity bit fails, the Validity bit is reset to indicate possible corruption in the sub-frame data.

C (Channel status) transmits information about the signal itself and U (User bit) can, as the name implies, be used by the user. C and U are collected on a running basis into 8-bit bytes.

FRAMES AND BLOCKS

Two sub-frames together comprise an entire frame of 64 bits $(2 \cdot 32)$ in total. In audio, this will normally be one sub-frame with data from the left channel and one sub-frame with data from the right channel. If the sampling frequency is for example 48 kHz, then it means that 64 bits must be transferred 48,000 times per second. This results in a bit stream of a good 3 Mbps (megabits per second). In connection with the expansion of the standard for use with higher rate sampling (88.2 or 96 kHz), two sub-frames can be used to contain samples from a single channel.

A block is formed from 192 pairs of samples or frames. At a sampling frequency of 48 kHz this results in 250 blocks per second. A byte is comprised of 8 bits. In each block, $2 \cdot 24$ $(2 \cdot (192/8))$ channel status bytes, and the same number of user-defined bytes, can be formed by left channel user data and right channel user data, respectively. Sub-frame $0-7$ forms the first byte, $8-15$ forms byte number two, etc. Each byte in a block has its own significance. With 250 blocks and thus that many bytes per second (at $f_s = 48$ kHz), a great deal of extra information can be transferred along with the sound.

CHANNEL STATUS

The information that is contained in channel status can be used to indicate what the actual bit stream consists of and where the individual sample belongs, and it can also provide a full time code. The data contained may be protected with its own error correction code (CRC).

As mentioned earlier, 24 bytes can be established within one block. Of these, we will only look at the first five. This information is defined in the

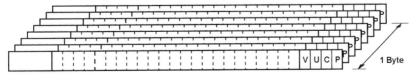

FIGURE 23.2 AES: For every 8 sub-frames, one byte can be formed. One block contains 192 frames $(2 \cdot 192$ sub-frames$)$.

AES3 standard. While the AES3 standard is the most comprehensive protocol, other interface protocols use similar information.

Status Byte 0

It is here the most significant difference is found between the professional and consumer versions. Depending on the value of the first bit, the channel status will have a different meaning. The AES deals in principle only with the professional version. The consumer information lies in the protocol for IEC 60958, also known as S/PDIF.

TABLE 23.1 AES3, status byte 0.

	Bit	State	Meaning
Use of channel status block	0	0	Consumer use of channel status block.
		1	Professional use of channel status block.
Linear PCM identification	1	0	Audio sample word (ASW) represents linear PCM samples.
		1	ASW used for purposes other than linear PCM samples.
Audio signal emphasis	2 3 4	000	No emphasis indicated, the receiver will make the decision.
		100	No emphasis, manual selection not possible.
		110	Emphasis 50/15 μs. Receiver manual override is disabled
		111	Emphasis ITU J.17 (with 6.5 dB insertion loss at 800 Hz). Receiver manual override is disabled.
		001	Reserved for later use.
		010	Reserved for later use.
		011	Reserved for later use.
		101	Reserved for later use.
Lock indication	5	0	Sampling rate lock not indicated.
		1	Sampling rate of source not locked.
Sampling frequency	6 7	00	Sampling rate is not indicated.
		01	Sampling rate 48 kHz.
		10	Sampling rate 44.1 kHz.
		11	Sampling rate 32 kHz.

Status Byte 1

This byte contains information on the relationship between the audio signals in the two channels. After higher-rate sampling became an option, an indication was also added for the relationship between sub-frames.

TABLE 23.2 AES3, status byte 1.

	Bit	State	Meaning
Channel mode	0 1 2 3	0000	Channel mode not indicated. Receiver default to two-channel mode. Manual override is enabled.
		0001	Two channel. Manual override is disabled.
		0010	Single-channel mode (mono). Manual override is disabled.
		0011	Primary/secondary mode (sub-frame 1 is primary). Manual override is disabled.
		0100	Stereo, channel 1 is the left channel. Manual override is disabled.
		0101	Reserved for user-defined applications.
		0110	Reserved for user-defined applications.
		0111	Single channel, double sampling frequency mode (first and second sub-frame share a sample).
		1000	Single channel – stereo: left – double sampling frequency mode (first and second sub-frame share a sample).
		1001	Single channel – stereo: right – double sampling frequency mode (first and second sub-frame share a sample).
		1111	Multi-channel mode, vector to byte 3 for channel identification.
User bit management	4 5 6 7	0000	Default, no user information is indicated.
		0001	192-bit block structure with user-defined content. Block start aligned with channel status start.
		0010	Reserved for the AES18 standard.
		0011	User defined.
		0101	192-bit block structure as specified in AES52. Block start aligned with channel status start.
		0110	Reserved for IEC 62537.
		0111	
		|	Reserved for future use.
		1111	

Status Byte 2

Here information is found on auxiliary bits, word length, and alignment level.

TABLE 23.3 AES3, status byte 2.

	Bit	State	Meaning	
Use of AUX bits	0 1 2	000	Maximum audio sample word length is 20 bits (default). Use of AUX bit is not defined.	
		001	Maximum audio sample word length is 24 bits (default). AUX bits are used for main audio samples.	
		010	Maximum audio sample word length is 20 bits. AUX bits in this channel are used to carry a single coordination signal.	
		|011	Reserved for user defined applications.	
		100		
		|	For later definition	
		111		
Encoded audio sample word length of transmitted signal			Audio sample word length if maximum length is 24 bits as indicated by bits 0 and 2 above.	Audio sample word length if maximum length is 20 bits as indicated by bits 0 and 2 above.
	3 4 5	000	Word length not indicated (default).	Word length not indicated (default).
		001	23 bits	19 bits
		010	22 bits	18 bits
		011	21 bits	17 bits
		100	20 bits	16 bits
		101	24 bits	20 bits
		110	For later definition	For later definition
		111	For later definition	For later definition
Indication of alignment level	6 7	00	Alignment level not indicated	
		01	Alignment level SMPTE RP155 (20 dB below max code)	
		10	Alignment level EBU R68 (18.06 dB below max code)	
		11	Reserved for future use	

Status Byte 3

In multi-channel mode this byte can indicate which channel is involved.

Status Byte 4

The last part of this byte was put into use for the indication of sampling frequencies that were introduced after the original standard ones.

TABLE 23.4 AES3, status byte 4.

	Bit	State	Meaning
Digital audio reference signal	1 0	00	Not a reference signal.
		01	Grade 1 reference signal (ref AES11).
		10	Grade 2 reference signal (ref AES11).
		11	Reserved and not to be used until further defined.
Information hidden in PCM signal	2	0	Not indicated (default).
		1	Audio sample word contains additional information in the least significant bits (ref AES55).
Sampling frequency	3 4 5 6	0000	Not indicated (normal).
		0001	24 kHz
		0010	96 kHz
		0011	192 kHz
		0100	Reserved
		|	Reserved
		1000	Reserved
		1001	22.05 kHz
		1010	88.2 kHz
		1011	176.4 kHz
		1100	Reserved
		1101	Reserved
		1110	Reserved
		1111	User defined
Sampling frequency scaling flag	7	0	No sampling frequency scaling.
		1	Sampling frequency is 1/1.001 times the value in bits 3–6 or 6–7.

AES3 AS A GENERAL FORMAT

Regarded as an interface, AES3 can also be used as a general transport format, for example for bit-reduced multi-channel sound. The interface is used for example for Dolby E, which is a standard for bit-reduced sound in six channels with accompanying metadata. Similar possibilities exist for MPEG formats.

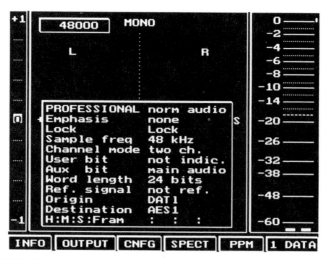

FIGURE 23.3 Significant information gathered from channel status and made available on the screen of the instrument. Among other things, one can derive that this concerns professional/two-channel audio/48 kHz/24 bit/without emphasis/no time code.

BI-PHASE MODULATION

Before the digital bit stream in the AES3 signal is transmitted, it is subjected to bi-phase modulation. This is the same coding as is used in SMPTE time codes. With bi-phase modulation, a shift is performed (from 0 to 1 or vice versa) for each new bit in the signal. In addition, a shift occurs each time the bit value is "1." The advantages of bi-phase modulation include that the signal is free of DC, independent of connecting polarity, and in general self-clocking.

IEC 60958-3 — 2006 CONSUMER APPLICATIONS

IEC 60958-3, formerly known as the Sony Philips Digital Interface, or just S/PDIF, was defined as a consumer format. The structure of the data stream is the same as in AES3. There are certain differences in that the format also contains the transmission of four-channel usage. Another significant point where it differs is copy protection (copy-prohibited).

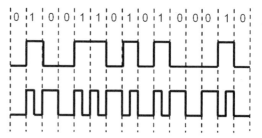

FIGURE 23.4 Bi-phase modulation. Top: Digital information. Bottom: Same signal, but bi-phase modulated.

The physical connection deviates from AES3, in that a 75 ohm unbalanced coax cable with RCA-phono jacks is used. The electrical level ought to be above 1 Vpp; however, values down to 400 mVpp can be read. In practice, this means that there are abundant opportunities for S/PDIF and AES3 to be able to understand each other. S/PDIF can also be run on fiber optic cables.

Channel Status

The structure of the bit format is, as mentioned earlier, comparable with AES3. The difference is indicated in channel status byte 0. If the value is 0, then the subsequent bits and bytes in channel status will have meanings that differ from AES3.

Status Byte 0

The general information on the audio format is placed here.

TABLE 23.5 IEC 60958-3, status byte 0.

	Bit	State	Meaning
Use of channel status block	0	0	Consumer
		1	Professional
Linear PCM identification	1	0	Audio state
		1	Data state
Copyright	2	0	Copyright
		1	Non-copyright
Emphasis of audio	3	0	No emphasis
		1	50/15 μs emphasis
No. of channels	5	0	2-channel audio
		1	4-channel audio
	6-7	00	Reserved
		01	Reserved
		10	Reserved
		11	Reserved

Status Byte 1

This byte is reserved for information on the type of device that generated the signal. The original categories encompassed: general, CD, PCM adapter, or DAT. The specifications have later been extended in step with the widespread use of digital media.

TABLE 23.6 IEC 60958-3, status byte 1.

Value	Meaning
100XXXXL	Laser/optical devices
010XXXXL	Signal processors
110XXXXL	Magnetic tape or disk
001XXXXL	Digital Audio Broadcast
101XXXXL	Musical instruments, microphones, and other types of digital sources

Status Byte 2

TABLE 23.7 IEC 60958-3, status byte 0

	Bit	State	Meaning
Source		0000	Source No. without significance
		1000	0–3
		0100	Source No. 2
		1100	Source No. 3
			etc. up to
		1111	Source No. 15
Channel No.	4 5 6 7	0000	Channel No. without significance
		1000	Channel A (left if stereo)
		0100	Channel B (right if stereo)
		1100	Channel No. C
			etc. up to
		1111	Channel No. O

OTHER INTERFACE STANDARDS

AES3 has been the basis for special applications. AES42 is a standard for digital microphones. It is very much like AES3; however, metadata bytes are defined for signaling the make, type, and settings, etc. In the SDI (serial digital video interface) four AES3 pairs are embedded for audio purposes.

There are of course other important interfaces besides AES3 and the standards closely related to it. The formats can differ from each other by their primary purpose, the number of channels, the quantity of metadata, etc.

Bibliography

AES3-1-2009: AES standard for digital audio − Digital input-output interfacing − Serial transmission format for two-channel linearly-represented digital audio data − Part 1: Audio Content.

AES3-2-2009: AES standard for digital audio − Digital input-output interfacing − Serial transmission format for two-channel linearly-represented digital audio data − Part 2: Metadata and Subcode.

AES3-3-2009: AES standard for digital audio − Digital input-output interfacing − Serial transmission format for two-channel linearly-represented digital audio data − Part 3: Transport.

AES3-4-2009: AES standard for digital audio − AES standard for digital audio − Digital input-output interfacing − Serial transmission format for two-channel linearly-represented digital audio data − Part 4: Physical and electrical.

AES42−2006: AES standard for acoustics − Digital interface for microphones.

AES47−2006: AES standard for digital audio − Digital input-output interfacing − Transmission of digital audio over asynchronous transfer mode (ATM) networks.

AES47-Am1-2008: Amendment 1 to AES47 − AES standard for digital audio − Digital input-output interfacing − Transmission of digital audio over asynchronous transfer mode (ATM) networks.

AES50 − 2005: AES standard for digital audio engineering − High-resolution multi-channel audio interconnection.

AES51 − 2006: AES standard for digital audio − Digital input-output interfacing − Transmission of ATM cells over Ethernet physical layer.

AES58 − 2008: AES standard for digital audio − Audio applications of networks − Application of IEC 61883-6 32-bit generic data.

AES-10id-2005: AES information document for digital audio engineering − Engineering guidelines for the multichannel audio digital interface, AES10 (MADI).

Where to Connect a Meter

CHAPTER OUTLINE

A level meter can be used at various points of the signal chain.

ANALOG CONNECTION

Before selecting points where the instrument is to be inserted, it is worthwhile to look at a couple of fundamental concepts concerning measurement technique. One must know whether the connection of the measurement instrument will have an influence on the measurement itself. The keywords here are **voltage matching** and **impedance matching**.

VOLTAGE MATCHING

A basic electronic circuit showing the output of a device connected to another device (a level measuring instrument in this case) is displayed in Figure 24.1.

Audio Metering. DOI: 10.1016/B978-0-240-81467-4.10024-3

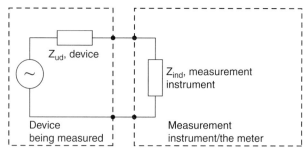

FIGURE 24.1 The connection between two devices can be viewed as a division of voltage between the source impedance (Z_{out}) and the input impedance (Z_{in}).

The rule is that the largest voltage is found across the largest resistance or impedance. If measuring the magnitude of the output signal of the "device," then it is necessary that the measurement system not load the output. Therefore, the input impedance must be high compared to the source impedance. A good rule of thumb says that the input impedance in the measurement system must be at least 10 times that of the output impedance of the device that is being measured. This applies as long as we are dealing with LF (low frequency), i.e., when we are in the audio spectrum below 20 kHz and the cable lengths are less than a few hundred meters/yards.

A typical example: The output impedance of a mixer's line output or aux-send can be 100–200 Ω. The input impedance of the analog input of a measurement instrument (in this case from DK-Technologies) is specified to be >20 kΩ. The input impedance is at least 100 times higher than the output impedance; thus the measurement instrument will display the correct signal voltage on the output concerned.

IMPEDANCE MATCHING

When dealing with LF in cables with lengths in kilometers/miles and for HF (high frequency), for example when handling video signals, antenna signals, digital audio signals, etc., there is a need for impedance matching. Impedance matching means that the source impedance and the load impedance must be equal. If there is an impedance mismatch, reflections and signal degradation will occur in the cable.

The consequence of impedance matching is that the signal becomes equally divided between the two impedances in terms of voltage. In other words, there is half of the voltage over each. In order to measure the correct level with a measurement instrument with high input impedance, the output that requires impedance matching must be terminated correctly, i.e., loaded with the nominal impedance. If this is not the case, the measurement instrument will show a value that is up to 6 dB too high. If, for example, an output is specified +4 dBm, it refers to the fact that the signal is 1.23 V, although generally only when the output is terminated with 600 Ω.

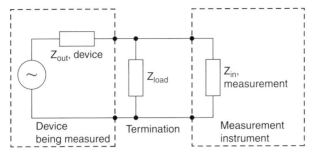

FIGURE 24.2 Impedance matching: The level is measured correctly when the output is terminated (loaded) with the nominal impedance. Without loading, double the voltage will be measured when the measurement instrument exhibits high input impedance.

THE MIXING DESK

The need for level measurement arises first and foremost when controlling levels with the sound mixing desk.

Inputs   The signals in the inputs are at a microphone or line level. Here, an overloading indicator is of primary importance. Each channel must be driven optimally, but without overloading.

Master outputs (main out)   Here, it will always be natural to measure the levels relating to the driving of the downstream step: recording media, transmitter, PA system, etc.

Group outputs (group out), aux and bus outputs   Larger audio mixers will normally either provide meters on these outputs or include the option to allocate one or more instruments for this purpose.

Insert points and aux-return   Here, it will be less common for a meter option to be built into the mixer. It can therefore be beneficial to have the instrument(s) connected to a patch panel, so that it is possible at any time to monitor whether the return channels are also appropriately driven.

There can be a difference between the nominal levels at different places in the mixer. Because the level at the insert points may be lower than the group and master outputs, it is a good idea to be familiar with the mixer's level diagram.

OUTPUTS WITH EMPHASIS

Often the level meter will be used to measure a signal that will subsequently be emphasized, i.e., corrected in certain parts of the spectrum. One example involves the output from a radio station that will subsequently be connected to an FM transmitter. Here, pre-emphasis is performed by raising the treble. Upon reception, the treble is lowered correspondingly. A form of noise reduction is attained by doing this (see the discussion on emphasis curves in Chapter 9). A second example involves an audio signal that is transmitted by satellite, where the J.17 standard is used. Emphasis can even be involved in the use of digital media. AES/EBU includes the possibility for signaling using two different standards, and S/PDIF includes the possibility for one.

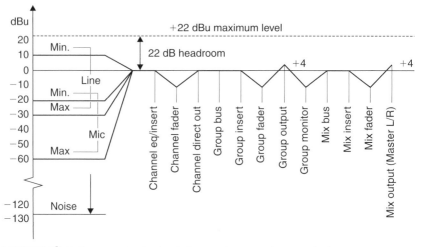

FIGURE 24.3 An example of a level diagram for an analog sound mixer.

Most emphasis curves were developed when it was not particularly common to have extreme modulation in the treble. Today, the problem can be that, even though a great deal of care has been taken in keeping the level under the applicable limits, the emphasis can still cause the transmission line to be overloaded, precisely because of the raising of the treble. Quite literally, one can risk burning the FM transmitter out or having the satellite transmission drop out.

In order to protect against accidents it is thus not just an advantage, but rather a necessity to be able to connect an instrument after the emphasis. In the same regard, this also holds true for the program limiter, i.e. the limiter that normally sits on the output as protection against overloading.

BALANCED/UNBALANCED

In order to run an electrical signal from one device to another, two wires must be used. Otherwise a circuit is not established. In a balanced connection, these conductors are physically identical. The shielding around the conductors can comprise a third wire though this is not included in the signal path. In an unbalanced connection, there is one conductor involved that is "hot" and the shielding is then used as the other conductor in the circuit. The shielding and hence the chassis are part of the signal path.

When the meter is connected to the output it can occur in some instances that the reading is not what one expected. On the following pages different situations are described.

Unbalanced Out/Unbalanced In

In an unbalanced connection, the correct level is in principle always being measured. There can, however, be unfortunate situations in which there might be problems with low-frequency noise (hum) if both systems are connected

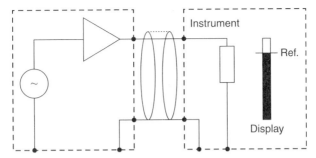

FIGURE 24.4 Unbalanced output; unbalanced input.

to ground via, for example, the power plug. A low-frequency noise loop can arise in the earth/frame/shielding circuit.

Unbalanced Out/Balanced In

The connection can be problematic if the pins of the connector have not been configured correctly. The aim is of course to establish a circuit. After this the concern is to preserve the balancing on the wires where possible.

If both leads of the balanced connection are not connected, then no circuit is established and the instrument will not display anything.

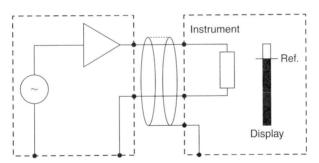

FIGURE 24.5 Unbalanced output; balanced input. A circuit is established. The display is OK. Note that the shielding is only connected to the receiving device.

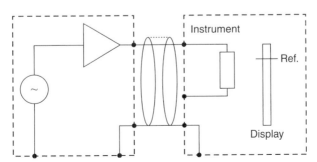

FIGURE 24.6 Unbalanced output; balanced input. A circuit has not been established. Hence nothing is displayed.

Balanced Out/Unbalanced In

With a balanced output connected to an unbalanced input, there are several possible errors that relate to how the balanced output is established. For electronic balancing, there are different possible circuits. The balancing may also be attained by using a transformer.

The main problem is the electronic balancing as shown in Figure 24.7. However, if a signal must go through to an unbalanced input, then the connection must be implemented as shown. The two frames must not be connected to each other.

In Figure 24.8, the connection goes totally wrong because the signal is grounded before it reaches the inverter.

Figure 24.9 shows a more common type of electronic balancing.

The following connection (Figure 24.10) is dangerous because there is a risk of only measuring half the signal, which will happen if one lead is not connected.

When an output is balanced by a transformer (Figure 24.11), the circuit is somewhat more manageable.

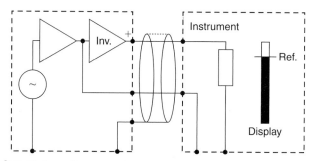

FIGURE 24.7 "Balanced" output; unbalanced input. The output is not correctly balanced, but the signal can be measured as long as the two chassis are not connected together.

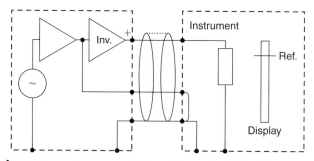

FIGURE 24.8 "Balanced" output; unbalanced input. No display! Because of the poor design of the output circuit, the signal is grounded due to connection to the chassis.

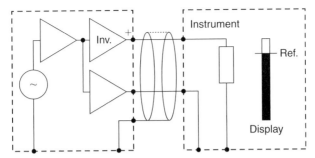

FIGURE 24.9 "Balanced" output; unbalanced input. The circuit is established, and the display is OK. However, the balancing is referenced to the chassis.

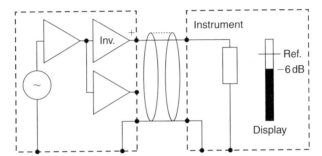

FIGURE 24.10 "Balanced" output; unbalanced input. Because the output is still referenced to the chassis, a signal arrives at the output, but only half of it (−6 dB)!

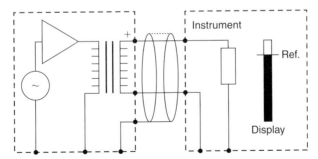

FIGURE 24.11 Balanced output; unbalanced input.

Balanced Out/Balanced In

When both the input and output are balanced, there is little that can go wrong. The circuit is created independent of whether the shielding is run through or not. It is then possible to consider possible ground and shielding strategies without thinking about the signal itself.

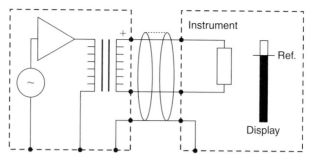

FIGURE 24.12 Balanced output; balanced input. A circuit is created regardless of whether the shielding is run through or not. The display is OK.

CONNECTIONS VIA A JACKFIELD (PATCHBAY)

If you are in possession of an installation with a jackfield (patchbay), it is always practical to have access to an instrument via that route. The main instrument can be run through the jackfield (patchbay) via so-called "half normalization" (i.e., the normal signal to the instrument is broken when a connection is inserted from the output one wishes to check).

LIVE SOUND

Regardless of whether you are producing live sound on a PA system or recording "live to tape" it is necessary to monitor the sound. This applies in particular to critical sources such as wireless microphones, or sources that are extremely variable in level.

It is not just the levels that have to be kept under control, but also the phase relationship between two outputs in a stereo setup; particularly if the signal will be played back in mono. A goniometer is almost a must-have in this situation.

THE INSTRUMENT AS A MEASUREMENT DEVICE

When you are in possession of a well-calibrated instrument, it is possible to use it as a measuring instrument. As a rule, an instrument that displays RMS can show the level of all signal voltages in the audio spectrum, particularly if it has an appropriately high input impedance (>10 kΩ).

One must know the significance of the scale, for example whether it is calibrated in dBu, such as with the Nordic scale on a PPM instrument. A level of 0 dBu corresponds to 0.775 volts. Many instruments have the option for amplifying the input signal, typically by 20 dB. 20 dB corresponds to an amplification of precisely 10 times. With the instrument's logarithmic scale, it hence becomes possible to measure signals at less than 1/1000 of the instrument's full scale. If the instrument ultimately will be used for larger signals, then an attenuator (of, for example, 20 or 40 dB) can be acquired. In this manner it also becomes possible to perform measurements in installations such as 100-volt loudspeaker systems.

DIGITAL CONNECTION

If you are in possession of a digital instrument, then there is the possibility of monitoring a number of functions at the same time. It is the instrument that the eye first falls on when the digital signals go haywire. A digital instrument is as a rule equipped with an AES/EBU or S/PDIF input. However, other interfaces are available as well.

The Meter is the Slave Device

When connecting it, the instrument will be designated as a slave. In other words, it will be clocked from the equipment that it is connected to. The AES/EBU interface is self-clocking, so if the meter is connected, then the instrument can follow along automatically.

It sounds simple, but it can be a problem if you have to use your mixer's only digital output for the instrument, meaning that the output signal has to be analog. That of course is not good. Therefore, it can be quite practical to be in possession of a router or a switch, giving the option to make a connection as required. In many smaller studios, however, this is not the type of thing that money is used on first, so other solutions may have to be found.

One of the more economical possibilities is that you can place your meter on the output of the master machine that you are recording on, since it would be connected as a slave to the mixer. Another inexpensive variation could be to connect the meter to an S/PDIF coax output. It ought to be possible to get a digital meter to read the interface, even though the voltage is a little lower than AES/EBU. Certain information will be lost; however, you still have the information on level, on the sampling frequency, on the number of bits per sample, and on whether the units are locked to each other.

Sample Rate Converter

In everyday terms, the many sampling frequencies that are worked with in practice can certainly be a bit cumbersome. Because of this, a sample rate converter is frequently available. This allows all arbitrary digital signals to be converted to a common standard.

Where the digital instrument gets its signal from must be known. Some converters are not transparent and thus not all data come through, even if only the sampling frequency was changed.

Clock Distribution

The ideal for a setup with multiple digital devices connected to one another is to have a distributed central clock. It is not guaranteed that the level meter will have an input for an external clock; however, if the rest of the system runs on the distributed clock, then it can be significantly easier to connect the meter as required.

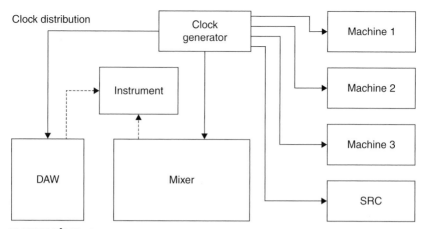

FIGURE 24.13 Setup of digital equipment with clock distribution. There is a clock generator that is shared by as many devices as possible, such as a mixer, digital audio workstation (DAW), and sample rate converter (SRC).

Networking

Analog audio needs a separate physical circuit for each channel. With digital audio, many channels can be carried in one circuit. The digital interface AES10 (MADI) can carry up to 64 channels. In digital audio networking it is possible not only to transport audio data but also to address the individual units attached to the network and to circulate lots of control data. A number of proprietary networks exist for different purposes, some of them based on Ethernet and the like. However, gradually general purpose open source networks like OSI, the Open System Interconnection model created by ISO, will find their way to the market.

When connecting to a network the meter, or the device including a meter, can be attached at any point in the chain.

LOUDSPEAKER/HEADPHONE MONITORING

Monitoring audio is of course not only a question of using visual displays. Listening is just as important — at least! So it is very practical to establish a relation between reading and the listening level. Further, it is important that the monitoring levels of speakers or headphones are well defined in order to mix for the right balance and keep levels from damaging hearing.

Regardless the type of signal distributed, analog or digital, a monitor controller with a built-in meter that is set to a reference listening level is an excellent device in all installations. Experience shows that less attention has to be paid to the visual meter when the engineer or editor is confident with the listening level.

FIGURE 24.14 An example of a combined D-A converter, meter, and monitor control unit for loudspeakers and headphones. Digital inputs: S/PDIF, AES3, TOS and ADAT (including the ability to confirm if those inputs are synchronous or not). (TC Electronic BMC-2)

Fast Fourier Transformation

CHAPTER OUTLINE

FFT is an acronym for "Fast Fourier Transformation." The purpose of a Fourier Transformation is to transform waveform (impulse response/waveform) into frequency. This is a common methodology which is implemented in many kinds of audio software.

PERIODIC SIGNALS

The basis for the Fourier transformation is the fact that every (infinite) periodic signal can be described by an infinite harmonics series (cf. Chapter 4, "Signal Types").

The signal can be resolved into a set of constituents based upon the fundamental frequency. Each of these constituents contains information on both amplitude and phase. The periodic signal has a discrete frequency spectrum, where the distance between each of the lines in the spectrum corresponds to the fundamental frequency.

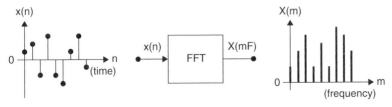

FIGURE 25.1 Principles of spectrum analysis performed using FFT.

Audio Metering. DOI: 10.1016/B978-0-240-81467-4.10025-5

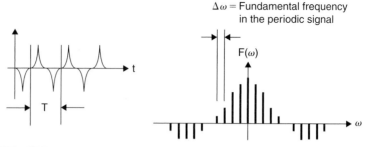

FIGURE 25.2 Left: Periodic signal. Right: Discrete spectrum of the periodic signal. The frequency distance ($\Delta\omega$) is then 1/T, where T is the period for the fundamental frequency.

However, not all signals are periodic. Nonperiodic signals do not repeat themselves. One can, however, establish the convention that the periodic signal constitutes a single period within the space of time under consideration. As a basis, the space of time under consideration is infinitely large. This in turn leads to the distance between the frequency lines in the corresponding spectrum being infinitely small. It can also be expressed in this manner:

$$\mathbf{\Delta\omega} \; = \; 1/T$$

where
$\mathbf{\Delta\omega}$ = the fundamental frequency
$\mathbf{T}$ = period
Hence if $\mathbf{T} \rightarrow \infty \Rightarrow \mathbf{\Delta\omega} \rightarrow 0$

It thus no longer involves a discrete spectrum, but rather a continuous spectrum.

FIGURE 25.3 Left: Nonperiodic signal. Right: Continuous frequency spectrum for nonperiodic signal.

FFT ANALYSIS

FFT is in principle a sampled version of the continuous spectrum. What is most significant about FFT is that the calculation can be performed rapidly when, in each block of samples (also called a frame), 2^n samples are selected, where n is an integer. Thus the number of samples can be 2, 4, 8, 16, ..., 1024, 2048, etc.

The sampling frequency is of course crucial to the frequency resolution of the signal. Thus in practice a sampling frequency is chosen so the fundamental

TABLE 25.1 Relationship between the Choice of Sampling Frequency and the Number of Samples (FFT Size) for a Given Resolution in the Spectrum

Sampling frequency	Number of samples (FFT size)	Frequency resolution
44,100 Hz	2048	21.5 Hz
44,100 Hz	1024	43.0 Hz
22,050 Hz	2048	10.7 Hz
22,050 Hz	1024	21.5 Hz
11,025 Hz	2048	5.4 Hz
11,025 Hz	1024	10.7 Hz

sampling rule is adhered to (the rule that the sampling frequency must be at least two times the highest frequency that one desires to reproduce).

It is the combination of the sampling frequency (sampling rate) and the number of samples (FFT size) in a frame that determines the frequency resolution in the spectrum. The relationship is described in the following expression:

$$f_{res} = f_s/N$$

where

f_{res} = frequency resolution in Hz
f_s = sampling frequency in Hz
N = the number of samples (FFT size)

Some examples follow of the resultant frequency resolution, that is, how large a distance there is between the individual points in the spectrum.

In practice, this means that details in the spectrum between the relevant frequencies are not shown. This is called "picket fencing," i.e., it is equivalent to viewing the world through a picket fence where you can only see between the slats. In the spectrum, all frequencies can be represented in the lines displayed, but if a more detailed display is desired then a higher resolution must be used with fewer cycles between the values displayed. This is done by reducing the sampling frequency (but also the upper boundary frequency) or by increasing the number of samples per frame (FFT size).

In FFT analysis, there will always be the same distance in cycles between the individual lines in the spectrum. Since it is usually preferred to view a frequency scale with a logarithmic axis, this means that the lines of the spectrum are shown closer to each other at higher frequencies.

DATA WINDOW FUNCTIONS

When an analysis is performed of an extract of a signal of longer duration, errors will arise in the analysis if a "softening" is not performed of the transitions between the boundaries. One could call this a form of "fade in" and "fade out."

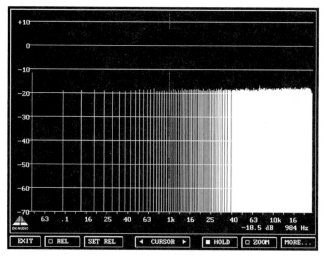

FIGURE 25.4 White noise is shown here in the form of an FFT spectrum. All lines are of the same height because white noise has constant energy per Hz. The lines are closer together at higher frequencies because the scale is logarithmic.

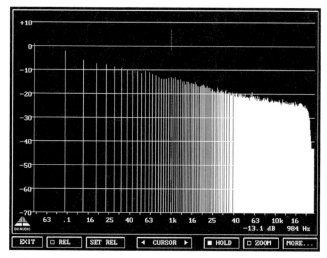

FIGURE 25.5 Pink noise is shown here in the form of an FFT spectrum. The height of the lines decreases towards higher frequencies because the energy in the signal is decreasing towards higher frequencies. However, the signal has constant energy per octave or per decade.

This is called a **data window function**. Each sample in the selected portion of the signal (each frame) is multiplied by a corresponding sample in the data window function.

The result of a "raw clip" is shown below in Figure 25.6. All samples in the data window function have the value 1. This is also called the **rectangular window**. It can be seen that the frequency function on the right includes a lot

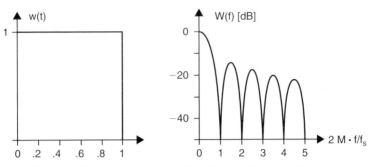

FIGURE 25.6 Left: Rectangular window. Right: The resultant amplitude spectrum with strong side lobes, the first at −14 dB.

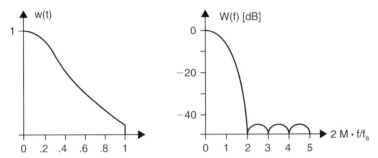

FIGURE 25.7 Left: The Hamming window. Right: The resulting amplitude spectrum with attenuated side lobes, the first at −43 dB.

of side lobes, which come into view as frequency components in the analysis performed, even though they are not present in the signal that is being measured. These frequency components arise when considering a section or extract of a continuous signal at full amplitude across the entire length of the selected section.

Other types of windows are thus calculated that can suppress these phenomena to an appropriate extent. Among them are: the triangular (Bartlett), Hanning, Hamming, Blackman, Blackman/Harris, Kaiser, Parzen, and Welch. Each of these types has its advantages.

If there are doubts about which data window function should be selected, then a good basic one is either the Hamming or Hanning. A Hamming window is shown in Figure 25.7. It should be noted that the individual side lobes have a far lower level here than was the case with the rectangular window.

OVERLAP

In certain analyses, in addition to the frame concerned, a little of the prior frame is also included. This is done for the sake of continuity. It is called **overlap**. The degree of overlap is specified as a percentage. For example, 50% overlap means that those samples that were located in the last half of the prior frame are

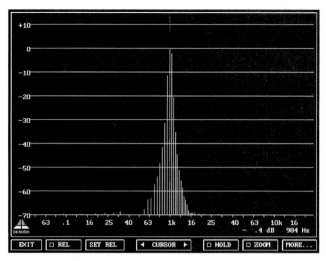

FIGURE 25.8 FFT analysis of a pure tone at 1 kHz modulated to 0 dB. The choice of the FFT parameters determines the appearance of the "skirts."

included in the current computations. Overlap is used in particular in analyses where a contiguous course of events is to be analyzed, for example spectrograms for the analysis of speech. Overlap is also used in the zoom function to provide greater resolution of low frequencies.

OTHER ANALYSES BASED ON FFT

Many analyses are, or can be, based upon FFT. Even though FFT is based upon linear frequency resolution, FFT is also used for logarithmic frequency divisions with a relative bandwidth, such as octave analyses. Here, the measured linear values are converted ensuring at the same time that there is a sufficient number of points available in each frequency band.

Certain systems (such as Meyer Sound SIM) use a full FFT within each octave.

INVERSE FFT

It is also possible to invert an FFT. This is called IFFT. A point of departure is taken here in the frequency content of the signal and the result is the signal's time function. This is used to implement digital filters in which the process first involves an FFT (from time to frequency). Once in the frequency domain, the signal processing operations are performed in which the spectrum is corrected. Then an IFFT is performed via which the signal is returned back to being a function of time.

DIGITAL FILTERS

With digital filters, it is possible to attain data that cannot be realized with analog devices. Among other things, these filters can be made phase linear in contrast to analog filters. It is also possible with digital technology to go much further, for example in the form of adaptive filters. Based on statistical methods, the coefficients of the filters can be determined on a running basis using the FFT analysis of the incoming spectrum.

BACKGROUND

Jean-Baptiste-Joseph Fourier (1768—1830) was a French mathematician who became one of the founders of mathematical physics. He developed the so-called Fourier series, which can be used for the mathematical treatment of periodic functions. The FFT algorithm was developed by John W. Tukey and James W. Cooley from IBM. It was originally developed to perform computations on stored data. As the operating speed of signal processors gradually increased, it also became possible to perform FFT analysis in real time.

Chapter | twenty-six

Spectrum Analyzer

CHAPTER OUTLINE

In the analysis of sound, it has always been important to be able to describe the frequency content of a signal. The traditional method is to use a filter bank, containing a number of filters with relative bandwidths, for example 1/1 octave, 1/2 octave, 1/3 octave, or perhaps with narrower bandwidths depending on the purpose. The signal being filtered passes through one filter at a time. Once the signal for a filter has been detected and its value read or printed, the next filter is reviewed.

A more practical variant is the parallel analyzer, where each filter has its own detector. With this, the values can be read simultaneously in all frequency ranges. A basic sketch is shown below of the parallel analyzer, in which the same signal is run to many filters simultaneously.

1/1 OR 1/3 OCTAVE

Filters with a relative bandwidth are used when what is being measured must be heard, since the ear also perceives frequencies logarithmically. In many control rooms and PA systems, it is normal to use analyzers based on 1/3 octave. By doing so, frequency responses can be measured for loudspeaker systems, etc.

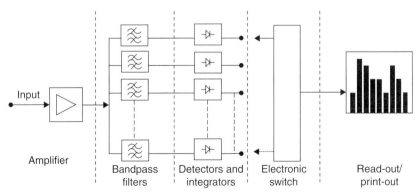

FIGURE 26.1 Principles of the parallel analyzer. The filters have a relative bandwidth, for example 1/3 octave.

Audio Metering. DOI: 10.1016/B978-0-240-81467-4.10026-7

201

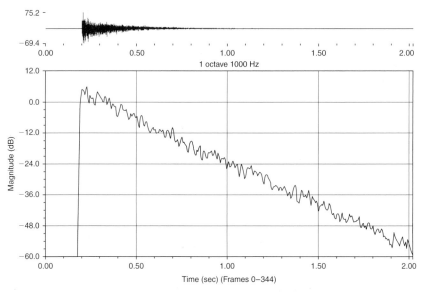

FIGURE 26.2 The measurement of a room's reverberation time can be performed by recording a shot in the room concerned. The decay is subsequently analyzed in octave bands. A time picture of a shot appears at the top. At the bottom, the decay is shown after the signal has been demodulated and made logarithmic.

In building acoustics, it is the case however that the absorption coefficients for materials are specified for each 1/1 octave. Hence it would be relevant to measure the reverberation time of a room in octaves. Alternatively, an average of the values in three 1/3 octaves can be calculated to obtain the octave value.

When measuring noise that is to be specified as noise rating (NR) or noise criterion (NC) curves, the analysis is always performed in 1/1 octave.

ANALOG OR DIGITAL

Now that digital technology has become common, analog filters by and large have not been made for measurements. Most of the filters used for these purposes today are based on FFT analysis, which is subsequently converted from a fixed to a relative bandwidth. It should also be mentioned though that information is not lost between the individual columns or "sticks" seen in FFT analysis. The areas between the frequencies are also taken into account.

One can, however, find octave or 1/3 octave analyzers that have two filter slopes to work with: one corresponds to the ISO standard for measurement filters, whereas the other represents the state of the art with much sharper filters that are obtained digitally as compared to standard filters defined in an analog manner.

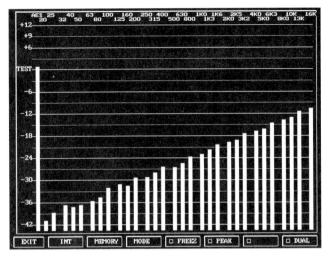

FIGURE 26.3 The spectrum for white noise analyzed in 1/3 octaves is shown here. White noise has constant energy per Hz. However, the filters have a constant relative bandwidth. This means that the individual frequency range covers more Hz the higher it lies in the frequency range. The column on the left shows the total level.

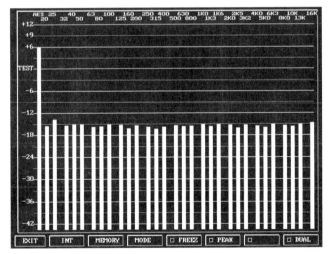

FIGURE 26.4 The spectrum for pink noise analyzed in 1/3 octaves is shown here. Pink noise has constant energy per octave; hence the horizontal curve made by the columns. The column on the left shows the total level.

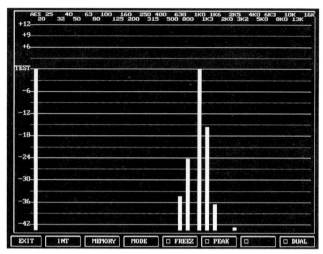

FIGURE 26.5 Analysis of a 1000 Hz sinusoid using a 1/3 octave analyzer. This bandwidth is obtained using a conversion from an FFT spectrum. Regardless of whether computed FFT filters or analog filters are involved, there will always be "skirts" on the analysis when pure tones are analyzed.

Chapter | twenty-seven

Other Measurement Systems

CHAPTER OUTLINE

Following are some very brief descriptions of several measurement systems/ technologies in addition to those already discussed.

TDS

Time Delay Spectrometry, or just TDS, is a technique that was originally developed by Richard Heyser of Jet Propulsion Lab.

The technique is based upon a known signal being transmitted, for example through a loudspeaker. The signal is typically a short frequency sweep, which is then recorded through the system via a microphone. The receiver can be synchronized in relation to the measurement signal transmitted. A tracking filter is synchronized so that it passes the frequency concerned when it arrives. When the sound is reflected and returned to the microphone shortly afterwards, the filter will be tuned on to a different frequency, and hence the reflected sound will be rejected. In this manner, a free field can be simulated without any influence from reflections. The filter can of course also be delayed so that the reflections can be measured on their own if needed.

(Reference and further reading: "Time Delay Spectrometry" edited by John R. Prohs, published by the Audio Engineering Society.)

What is interesting about the TDS system is that it can emulate a large number of other measurement systems and that it is in a position to measure everything from distortion to reverberation time and speech intelligibility in rooms.

Audio Metering. DOI: 10.1016/B978-0-240-81467-4.10027-9

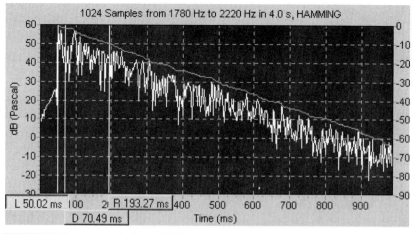

FIGURE 27.1 An analysis from a TDS measurement system such as those supplied by TEF®.

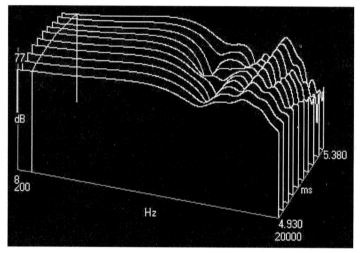

FIGURE 27.2 Waterfall plot with frequency spectra, generated by TEF®.

MLSSA/MLS

MLSSA is pronounced "Melissa" and is an acronym for Maximum-Length Sequence System Analyzer. This is a system for acoustical analysis of rooms and electro-acoustical components and systems. In contrast to other methods, MLSSA can be used for measurements over a large frequency range and a long time span at the same time.

MLSSA, as well as many other later developed systems, is based on the MLS (Maximum Length Sequence) method. The excitation signal is a periodic sequence of pseudorandom binary digits. A measurement system is typically based on a sound card in a computer. Measurements can be performed in a very short time and various post-processing options for the measurement data can give information concerning impulse response, the reverberation time in a room, frequency response, and speech intelligibility for electro-acoustic systems. MLS-based measurement systems can also emulate more traditional systems, such as the 1/1 octave or 1/3 octave analyzer.

SPECTROGRAPH

The spectrograph is an instrument — or software package — that can simultaneously display three parameters of a sound signal. It shows time on one axis (the x-axis) and frequency on another axis (the y-axis), and the level is shown by the degree of density. Some analog spectrographs are still in use. However, most spectrographs today are based on FFT analysis and run on computers.

The spectrograph is used in particular for voice analysis (both humans and animals), utilizing filters with both narrow and relatively large bandwidth. However, the spectrographic display (in colors) also forms a basic editing window for "spectral editing," where specific frequency components can be altered without affecting other parts of the frequency range.

Figures 27.3 and 27.4 show an analysis of the author's utterance: "Audio Metering." Different filter bandwidths have been applied in each figure.

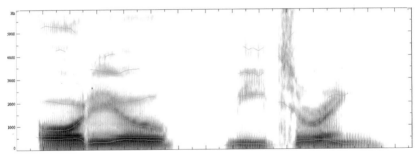

FIGURE 27.3 Spectrogram of the author's utterance: "Audio Metering." The horizontal axis is time, and the vertical axis is frequency (linear scale), whereas the degree of density indicates the level. A wide filter of 100 Hz was used, so that the individual harmonic overtones in the voice are not seen, but rather the formants, which are comprised of multiple harmonic overtones within a certain frequency range.

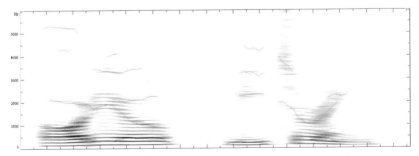

FIGURE 27.4 An analysis of the same sequence as above is shown here, but a narrow filter (10 Hz) has been used. It now shows a plot of the overtones instead. The analyses were performed with ST^X.

TRANSIENT ANALYSIS

A very large part of the current measurements and analyses of sound involve finding the RMS values of the signals, either in the entire spectrum or as a function of the frequency. The signals are regarded as constant within the period concerned.

Nevertheless, events can occur in the perceived acoustic image that are clearly audible, but difficult to measure using the traditional methods. Examples of this are things like a single click or "glitch" in the sound due to erroneous sampling, drop-out, or clipping. There can also be mechanical rattle from objects in the listening room, which are set in motion by the oscillations of the loudspeakers. Or by the loudspeakers themselves, which, due to the moving coils scraping against the magnet, produce some audible, but not easily measurable, noise. Transient analysis is the tool that makes a large number of these phenomena "visible." Instead of measuring the RMS value of the signal, the signal's instantaneous change in energy is measured.

This technology applies even to audio forensics when looking for possible edit points in the process of the authentication of digital audio recordings.

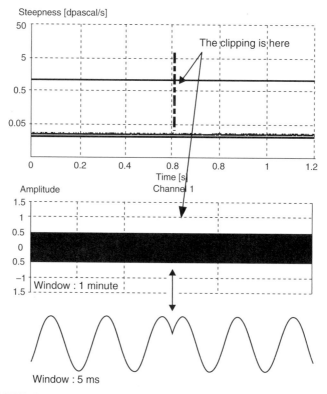

FIGURE 27.5 Example of a transient analysis of a sinusoidal tone where a clip has occurred. The analysis clearly shows where this audible click occurs, even though it cannot be seen on the time signal before zooming all the way in. The analysis was performed with Harmoni™ Lab.

SPEECH TRANSMISSION INDEX, STI

At the beginning of the 1970s, it was proposed that MFT, Modulation Transfer Function, be used as a method to describe the intelligibility of speech in a room. MFT expresses an apparent signal/noise ratio since not just background noise, but also the reverberation of the room, is regarded as noise in the signal transmitted. The method shows how "unscathed" a speech signal is, transferred from the sound transmitter to the listening position. MFT technology used for speech transmission has been given the name STI, short for Speech Transmission Index. The method is used both for the computation and the measurement of speech intelligibility.

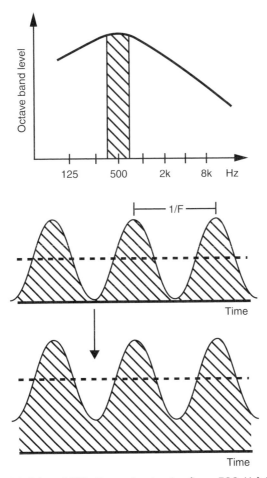

FIGURE 27.6 Principles of STI: Octave band noise (here 500 Hz) is modulated at a number of frequencies. When measuring STI, seven octaves are used, whereas for RASTI only two are used. In STI-PA all seven octaves are used however, with fewer modulation frequencies. The modulation index is determined for each modulation frequency and all results are combined to one single number.

This method uses the seven octave bands from 125 Hz to 8 kHz, which in total cover a frequency range corresponding to the spectrum of human speech. Each octave band is modulated by 14 different low frequencies, which in 1/3 octave steps go from 0.63 Hz to 12.5 Hz. The low-frequency modulation corresponds approximately to the modulation in speech. For each combination of carrier frequency (octave band noise) and modulation frequency, the MFT is computed or measured. This gives in all 98 sets of data, which are then reduced to a single number, that is the STI value for the combination of sound source and listening positions concerned in the room under investigation.

Subjective Scale

The STI values are related to a subjective scale, with designations running from "Bad" to "Excellent." An STI value of 0.6 or better is normally what is aimed for. This corresponds at least to "Good."

FIGURE 27.7 This scale indicates the relation between the STI-value and the subjectively assessed speech intelligibility

RASTI

Since a complete STI measurement is quite comprehensive and time-consuming, a reduced version called RASTI was developed. RASTI stands for RApid Speech Transmission Index. (Some call it Room Acoustic Speech Transmission Index.) Only the most significant combinations of carrier and modulation frequencies are included in it, amounting to nine combinations. However, the principles are otherwise the same as for STI. For a period of time, standardized RASTI measuring equipment has been the only practical way to perform measurements. The limiting factors of the RASTI measurement is that it does not accept nonlinearities in the signal chain like compressors, limiters, and the like, which in many cases may increase the intelligibility of a system. Therefore the RASTI values have primarily been valid for the measurements of pure acoustic spaces. The RASTI measurement is rarely used today, as a more comprehensive and yet easy-to-handle method has been developed.

STI-PA

STI-PA is short for Speech Transmission Index for Public Address Systems. Like RASTI, the STI-PA is a derivative of STI. However, this reduced methodology provides results that are sufficiently comprehensive to be compared to complete STI measurements. It was developed to cope with the nonlinear processing environment common to advanced sound systems, and to reduce the measurement time required to a practical level.

FIGURE 27.8 STI-PA readout from a handheld device (NTI AL1 Acoustilyzer). The display shows that the measurement is finished and that the value is 0.89 (which corresponds to "Excellent").

STI-PA supports fast and accurate tests with portable instruments that are able to evaluate speech intelligibility within 15 seconds per position. The STI-PA signal can be generated from a (professional quality) CD player or directly from a file-based signal generator. Implementing an artificial voice source an end-to-end system check including the microphone can be performed.

The procedure includes different routines to get the correct result: If the STI-PA value measured is less than 0.63, two more measurements must be made and all three results are averaged. If the results differ more than 0.03, three further measurements must be made and all six readings are averaged for the final results. Results that in the same position vary more than 0.05 should not be accepted. This kind of error may occur if the background noise is very impulsive or otherwise is varying. Measurements can be made at a time with less background noise and the results can be corrected using dedicated software.

Emulated STI Measurements

There are other types of measurement instruments, based on TDS and MLS, which are used to measure STI, RASTI, and STI-PA. Handheld microphones are basically not allowed. Additionally, time-varying background noise must be prohibited. Still, when comparing the results of the measurements from these systems, they are not completely in agreement with the "real" measurements and should not be used for documentation.

Standards

Following are some of the standards that require STI data:

ISO 7240 Fire detection and alarm systems.

NFPA 72 National Fire Alarm Code 2002

BS 5839-8 Fire detection and alarm systems for buildings. Code of practice for the design, installation, and servicing of voice alarm systems.

DIN 60849 System regulation with application regulation DIN VDE 0833-4.

LOGGING SYSTEMS

Noise regulations in the entertainment industry have been introduced in many countries. Even "awesome" music can be considered noise. Especially in venues for rhythmic music it is essential to control the sound level during a concert. The front of house (FOH) engineer should have proper information so he or she can keep the concert rolling and avoid being stopped by authorities due to excess of local limits.

In the measurement of environmental noise the limit is normally defined by the A-weighted equivalent level measured during a given interval of time ranging from minutes to hours. However, in order to control the level the time interval usually is in the range of 5—15 minutes.

Due to legislation, very often the levels measured have to be reported. Thus a noncorruptible logging system has to be installed.

Systems and programs for this special purpose have been developed. The best of these perform to the same standards as integrating sound level meters and include a calibrated microphone. Additionally, clear displays are provided showing the actual A-weighted sound pressure level and the C-weighted peak level. The A-weighted L_{eq} is calculated from the start. In some cases a special readout can tell if you are safe or if you are close to the limits and should "hold your horses." When the job is done an encrypted data file with logging data can be extracted from the system. Even if the connection to the microphone was broken during the measurement, this will be noted in the file.

European legislation is based on DIRECTIVE 2003/10/EC OF THE EURO-PEAN PARLIAMENT AND OF THE COUNCIL of 6 February 2003 on the minimum health and safety requirements regarding the exposure of workers to the risks arising from physical agents (noise).

FIGURE 27.9 Screenshot of readout from the measurement system "10eazy".

Chapter | twenty-eight

Measurement Signals

CHAPTER OUTLINE

The following is a short overview of some measurement signals that are useful to the sound engineer. Signals normally available are covered.

SINUSOIDAL TONES

The sinusoidal tone is practical because it can be kept constant and because it contains one (and only one) frequency. It is used for checking and calibrating equipment.

The sinusoidal tone is not practical for the calibration of acoustic sound levels (sound pressure) because the room influences the signal by virtue of reflections, etc. However, the specific problems of a room such as standing waves can easily be spotted using sinusoidal tones.

BURST

Tone bursts at well-defined intervals are the foundation for checking level-reading instruments. Either burst generators or CDs containing measurement signals can be used. Hard disk-based editing systems can also be used to generate the signals.

What is most important to remember is that the frequency must be relatively high (5−10 kHz) in order for the burst to contain sufficiently complete periods.

NOISE

Broadband noise signals, primarily in the form of pink noise, which has constant energy per octave, are indispensable in practical sound work. You can listen to the signal and you can use it together with spectrum analysis for adjusting loudspeaker setups, etc.

Noise can also be used in connection with reverberation measurements. Pink noise is fed to a loudspeaker placed in the corner of the room. The

Audio Metering. DOI: 10.1016/B978-0-240-81467-4.10028-0

decay that occurs when the noise is stopped can be recorded with an omnidirectional microphone. The microphone is moved to various positions and the resulting recordings are analyzed.

Pink noise generators may perform differently. Therefore, some standards for pink noise measurements are applicable and the crest factor of the noise is also defined. Typically the required crest factor is 4. This equals 12 dB.

CLICK GENERATORS

A click sound is a short impulse with a broadband spectrum. The click can be generated electrically and reproduced with a loudspeaker. It can also be a click that is generated acoustically by a clapper board or perhaps one of those "clickers" found in crackers under the Christmas tree. The latter are incredibly effective in examining coincidence in stereo microphone placements, etc. Be careful with your ears. The impulses are very short, but also very high (>120 dB re 20 μPa @ 10 cm typical). A click is also practical for examining echo phenomena in larger rooms.

An efficient form of controlled clicks is generated by spark generators. Dedicated spark generating devices are often used in connection with acoustic scale models, due to the contents of high frequencies in the spark (>100 kHz typical). The dynamic response of microphones can also be examined by use of spark sounds.

Phase checking systems often utilize unidirectional (positive or negative) clicks from an electric generator (short DC offset).

POP/BLAST

A somewhat more powerful impulse with a fair amount of low-frequency content is a balloon pop or perhaps a pistol shot (a blank) in order to examine large rooms and as a sound source in connection with reverberation measurements.

OTHER SIGNALS

Most measurement signals are related to dedicated measurement systems, as mentioned in Chapter 27.

Sound Level Meters

CHAPTER OUTLINE

Sound levels can be measured with a sound pressure meter, most often simply called a sound meter. A sound meter contains a microphone, amplifier circuit, and filter circuit, as well as a detector and display in the form of a moving coil type of instrument or a digital display. In addition, a sound meter can be equipped with some form of computational function, so calculations can be performed on sound levels recorded over a time period. There is thus a differentiation between sound pressure meters and integrating sound pressure meters.

MICROPHONE

A sound meter is equipped with a pressure microphone, i.e., a microphone with omnidirectional characteristics. Normally, a decision must be made concerning whether a microphone with flat free field or flat diffuse field characteristics is desired. For measurements outdoors, a wind cap is used in order to avoid wind noise in the microphone having an undesired effect on the results of the measurements. In order to be able to check the sound pressure meter's display, an acoustic calibrator is used. It is a small sound generator with a well-defined frequency and level.

AVERAGING

The detector rectifies the signal, averages it, and calculates an RMS value. There are two forms of averaging. One is linear averaging and the other is exponential averaging. With linear averaging, the RMS value is found for a fixed time interval that is called the integration time. It is this form of averaging that is used in level instruments. For example, the standardized

Audio Metering. DOI: 10.1016/B978-0-240-81467-4.10029-2

integration time is 5 ms for a PPM instrument according to IEC standard 60268-10.

In connection with the measurement of sound, exponential averaging is used. This means that the incoming signal is weighted so that the RMS signal "remembers the past," but events that lie further back in time will have less weight than events that have just occurred. The average time or the time constant is a measure of how fast the exponential function "decays," or more precisely, it specifies the time it takes before the exponential function has been reduced to 69% of the starting value. Internationally, the time constants used include 125 ms (called "F" for "Fast") and 1 second (called "S" for "Slow").

In integrating sound meters, measurements can be made over long periods of time (minutes/hours), and the result expresses the energy as that which a constant level would have had during the time period.

FREQUENCY WEIGHTING

Since the sensitivity of the ear is not the same at all frequencies, a filter can be inserted that takes this into account. The filter used most often performs an A-weighting, where the lowest and highest frequencies are attenuated, whereas frequencies in the 1–4 kHz range are amplified. Other filters are also used including B, C, D, and Z filters.

When, for example, an A filter has been used, the result of the sound pressure level read from the meter is written as xx dB(A) or as L_{pA}: xx dB. A sound meter can also be equipped with octave or 1/3 octave filters for use in the frequency analysis of the signal.

READING

The reading of the results is performed in dB. As a reference level for the measurement of sound pressure, the value 20 µPa $(2 \cdot 10^{-5}$ Pa) is always used.

OUTPUT

Normally, an AC output is available so the sound meter can function as a microphone pre-amplifier and so the signal can be recorded for later analysis. A DC output is often found too, so that a printer can record what the instrument's display is showing.

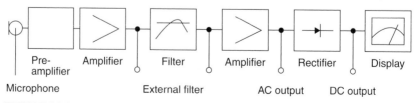

FIGURE 29.1 Basic diagram of the principles of a sound level meter.

CLASSES OF SOUND LEVEL METERS

The requirements for precision for sound pressure meters and for integrating sound pressure meters are given in the following two international standards:

IEC 60 651 Sound level meters

IEC 60 804 Integrating-averaging sound level meters.

The norms specify nominal values and tolerances for a number of characteristics:

Type 0 is intended as a laboratory reference standard.

Type 1 is intended for precise measurements in the laboratory and in the field.

Type 2 is envisioned for general measurements in the field.

Type 3 can only be used for a rough orientation concerning noise levels.

Type 0 has the narrowest tolerances. In most countries, it is required that measurements in the external environment must be performed with measurement equipment that adheres to the type 1 tolerances.

ANALYSES

In connection with the measurement of noise, different analyses are performed, including analyses of a more statistical nature. The most significant ones are listed below.

L_p SPL or Sound Pressure Level. The RMS value of the sound pressure concerned expressed in dB re 20 μPa.

L_{eq} The equivalent sound pressure level, i.e., the average value (on an energy basis) of the sound pressure level registered over a period of time.

L_E (previously called SEL). An expression of the total energy in the period being calculated.

L_n With a basis in the cumulative occurrence distribution, L_n specifies the sound level that has been exceeded in $n\%$ of the time interval under consideration. n can assume values from 1 to 99. L_n is often used to describe the background noise during periods with varying noise; for example, $L_{95} = 36$ dB means that this value was exceeded during 95% of the measurement period.

NR and NC Curves

CHAPTER OUTLINE

Noise rating (NR) curves and noise criterion (NC) curves are two sides of the same coin, only with different origins; hence NR is used in Europe, whereas NC is predominantly used in the North American countries.

NOISE RATING

Noise rating curves are a set of curves that are based on the sensitivity of the ear. They are used in order to arrive at a single number (rating) for noise, based on sound levels measured in octave bands (see Figure 30.3).

NR curves were originally developed as requirements curves for noise sources both indoors and outdoors. Now the curves are predominantly used in connection with requirements for, and documentation of, background noise indoors. Here, they particularly concern noise from ventilation systems and other similar noise from building installations. Up until 1972, NR curves were used as requirements curves for noise exposure at the workplace.

NR curves encompass the standard octaves from 31.5 Hz to 8 kHz and are defined in increments of 5 from NR-0 to NR-100. The NR curves are named after the value at 1 kHz.

The curves are used in the following manner. The noise is measured by a sound pressure with an attached octave filter. The measurement results are plotted on the sheet with the curve on it. Then the lowest of the NR curves that has not been exceeded is read off (see the example at the end of this chapter). This curve is then the result of the measurement (in the example shown it is called NR-25). It should be noted that NR values cannot be compared directly with dB(A) measurements. In the measurement and assessment of noise from ventilation systems, where the NR curves are often used, the A-weighted sound pressure level typically lies 4–9 dB above the NR value based on the measurement of the sound pressure level in octave bands. In this regard, note that the NR values are always specified or read off in increments of 5 dB.

Audio Metering. DOI: 10.1016/B978-0-240-81467-4.10030-9

NOISE CRITERION

Noise criterion (NC) curves are constructed in the same manner as NR curves, however only with the purpose of being used as requirement and measurement curves for indoor noise sources.

NC curves are also based on the sensitivity of the ear, only the values are a little different than the corresponding NR curves. In general, NC curves are a little flatter in comparison with NR curves, i.e., they have slightly lower values at low frequencies and slightly higher values at higher frequencies.

The curves were originally defined in octave bands running from 63 Hz to 8 kHz. In certain contexts, the curves are extrapolated down to the 31.5 Hz octave, as shown in Figure 30.1. In addition, extrapolation to 16 kHZ is used in that the value from 8 kHz is also used (this extrapolation is not shown on

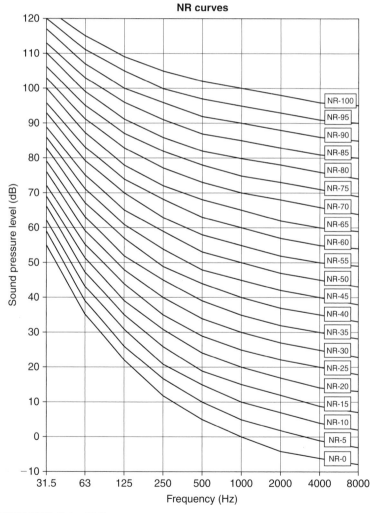

FIGURE 30.1 Noise Rating curves.

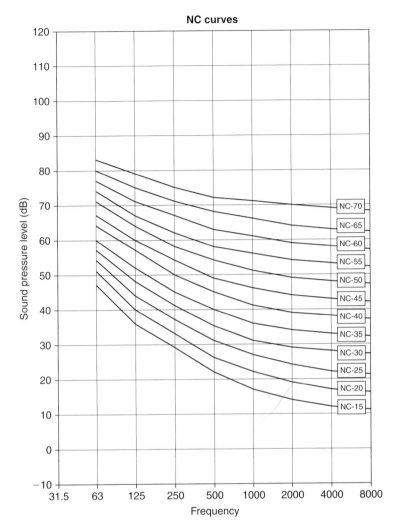

FIGURE 30.2 Noise Criterion curves.

the curve). The curves are defined in increments of 5, from NC-15 to NC-70. Measurements are made with the standardized "SLOW" integration time.

The reading takes place in the same manner as for NR curves: the octave levels are plotted and the lowest curve that is not exceeded is the result of the measurement (extrapolated curves are used).

NR REQUIREMENTS AND RECOMMENDATIONS

Recommendations for cinema audio, FSI 1992

    Class 2 cinemas, required: NR-30
    Class 1 cinemas, required: NR-25
    Class 1 cinemas, recommended: NR-20

If the noise contains audible tones, then the requirement is tightened by 5 dB.

NC REQUIREMENTS AND RECOMMENDATIONS

ISO 9568; 1993 (E): Cinematography: Dubbing and reviewing rooms, as well as first-run movie theaters: NC-25.

Example:

Noise is measured in octave bands, and the resulting NR value must be found. The following levels have been measured:

31.5 Hz:	*55 dB*
63 Hz:	*49 dB*
125 Hz:	*28 dB*
250 Hz:	*33 dB*
500 Hz:	*11 dB*
1 kHz:	*20 dB*
2 kHz:	*14 dB*
4 kHz:	*7 dB*
8 kHz:	*12 dB*

The levels measured are plotted on the data sheet with the curves (see Figure 30.3). The lowest NR curve that is not exceeded is NR-25. The octave band on the curve that comes the closest to the measured level is the 250 Hz octave. The measured noise thus corresponds to NR-25.

A measurement of the A-weighted sound pressure level for the same noise source would give a result of 29 dB(A).

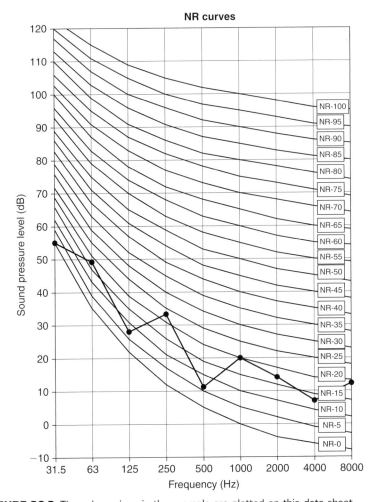

FIGURE 30.3 The values given in the example are plotted on this data sheet.

Room Acoustics Measures

CHAPTER OUTLINE

Room acoustics measures provide information about how sound behaves inside a room; how the sound is distributed and the influence of size, shape, surface materials, etc. Many of the measures are related to the way we perceive and assess the sound as listeners, as performers in the room, or as recording or SR engineers. In this chapter some basic room acoustics parameters and the related measures are briefly described.

In room acoustics in general we distinguish between small rooms like listening rooms, control rooms, recording booths, etc., and large rooms like concert halls, theaters, rock venues, etc. This is why different parameters to some degree are applicable to different room types.

Audio Metering. DOI: 10.1016/B978-0-240-81467-4.10031-0

GENERAL RULES FOR GOOD ACOUSTICS

When designing rooms for audio there is a general set of rules that can be reviewed to find a starting point to achieve "good" acoustics:

- appropriate reverberation time
- appropriate sound distribution
- low background noise
- no echoes (flutter echoes)
- appropriate control of early reflections.

While this set of rules is a good starting point for acoustics design in general, the solutions may turn out to be different depending on the application of the room. For instance, good sound distribution is vital in both an auditorium and a classroom. However, it is absolutely contrary to the requirement for open plan office environments with multiple workstation installations as found in many broadcast facilities these days.

SCHROEDER FREQUENCY

Before taking a closer look at the good rules we have to first consider the distribution of the sound waves within a room. When the physical size of the wavelength at lower frequencies approximates the dimensions of the room, the sound waves do not move freely. What we have are the so-called room modes, or eigenfrequencies. As with any other acoustical resonance phenomenon there is an emphasis on these frequencies that literally "fit" into the room. The result is an uneven distribution of the sound in this frequency range, the modal region. At higher frequencies, in the statistical region, the sound waves move freely and geometrically.

A practical approach to a definition of the transition between the modal region and the statistical region is the Schroeder frequency equation. The Schroeder frequency is defined in relation to the number of modes within a given frequency band. It is dependent on the volume of the room and reverberation time. However, this frequency is also dependent on the room shape (i.e., the sphere and the cube exhibit a higher Schroeder frequency compared to other shapes). The Schroeder frequency ($f_{\text{schroeder}}$) is defined as follows:

$$f_{\text{schroeder}} = 2000 \cdot \sqrt{\frac{T}{V}} [\text{Hz}] \text{ (unit: meter)}$$

or

$$f_{\text{schroeder}} = 11885 \cdot \sqrt{\frac{T}{V}} [\text{Hz}] \text{ (unit: foot)}$$

where
T = the reverberation time [s]
V = volume [in m^3 or ft^3]

It is not possible to describe this frequency exactly as the room shape may affect the frequency distribution. However, Figure 31.1 shows the relationship between Schroeder frequency, volume, and reverberation time:

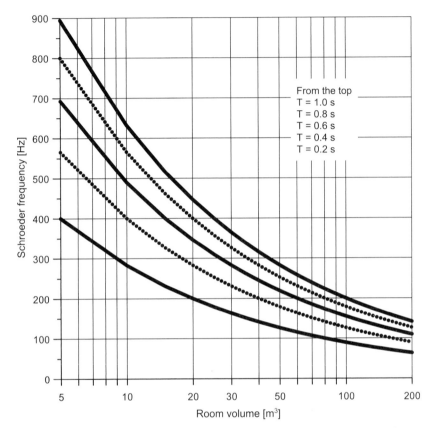

From the top
T = 1.0 s
T = 0.8 s
T = 0.6 s
T = 0.4 s
T = 0.2 s

FIGURE 31.1 Relationship between room volume and Schroeder frequency for T (reverberation time) = 0.2, 0.4, 0.6, 0.8 and 1 s.

REVERBERATION TIME

The reverberation time is the single most important parameter relating to room acoustics. Reverberation time is defined as the time it takes the sound field to attenuate by 60 dB after the sound source has stopped. Hence we also use the expression RT_{60} to represent reverberation time.

If a sound impulse is emitted from a sound source in a closed space, reflections bouncing from the surfaces can be observed. The more reflective (or less absorbing) the surfaces are, the longer it takes for the sound to die. The bigger the room, the longer the distance the sound must travel to meet either the reflecting or absorbing surface.

Figure 31.2(A) shows the impulse response of a room. In this case the sound impulse is actually the blast of a balloon. One can see on the waveform that

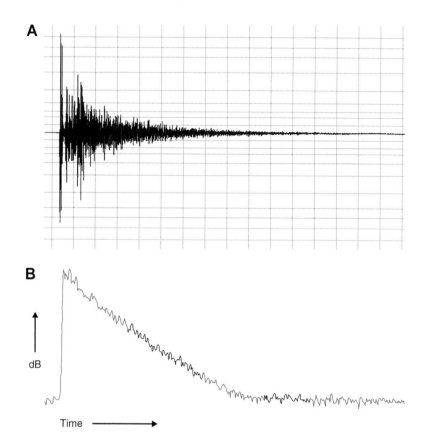

FIGURE 31.2   A: Impulse response in a room (waveform of a balloon blasted). All the individual reflections can be seen. B: The signal above has been converted to level (now the reading is in dB)

after the blast a lot of gradually diminishing impulses pass the recording microphone. When this response is converted to level (reading in dB) we can observe a linear decay (with minor bumps). It is the slope — measured in seconds per 60 dB — of this decay that expresses the reverberation time. However, we do not then necessarily need a 60 dB decay to find the reverberation time.

The RT_{60} (the reverberation time) is commonly defined by the part of the decay that has been observed or evaluated:

T_{30}: Slope in seconds per 60 dB evaluated over the −5 to −35 dB range of the decay curve.

T_{20}: Slope in seconds per 60 dB evaluated over the −5 to −25 dB range of the decay curve.

EDT (Early Decay Time): Slope in seconds per 60 dB evaluated over the 0 to −10 dB range of the decay curve.

T_{30} is used in general for analysis of the reverberation time. However, EDT is more related to the perceived reverberation time, especially if the slope

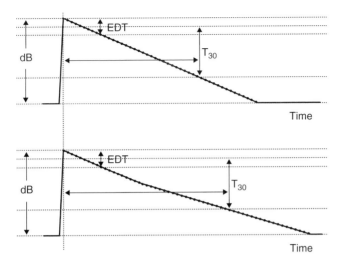

FIGURE 31.3 On the upper plot the decay is a straight line. On the lower, two different slopes can be observed. The EDT and T_{30} will have different values in the latter case.

sometimes changes at lower levels (which may occur in complex rooms). This will cause different values for EDT and T_{30}.

The reverberation time is in principle only relevant above the Schroeder frequency. However, we do try to measure reverberation below this frequency. The large variations that might be observed below the Schroeder frequency are compensated for to some degree by averaging measurements in many different positions of the room.

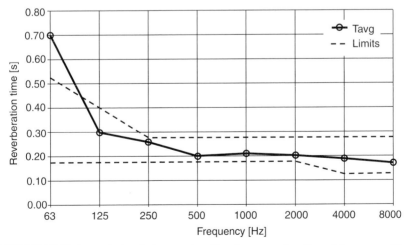

FIGURE 31.4 An analysis of the reverberation time as a function of frequency. The signal has been filtered in octave bands. In this example the reverberation time at lower frequencies (63 Hz octave band) exceeds the limits provided by a recommendation for small rooms for audio editing.

In general, the crucial information is how the reverberation time varies with frequency. Depending on the application, the reverberation time as a starting point should have the same value at all frequencies. In many cases it can become a major problem if the reverberation time at lower frequencies exceeds the midrange. Normally it is allowed that frequencies below 125 Hz may reach values of 1.5 times the midrange in studios, control rooms, and medium-sized rooms for music. However, in rock venues it is preferred that the low frequency reverberation is kept at the same level as the midrange.

When characterizing the reverberation time of a room with one single number, this is nominally the reverberation time around 500 Hz or in the midrange (i.e., 250—1000 Hz). Table 31.1 shows the generally preferred reverberation times at 500 Hz depending on the application of the rooms:

TABLE 31.1 Preferred reverberation times for different applications @ 500 Hz.

Application	Reverberation time	Comment
Vocal booth	0.1—0.2 sec.	The most difficult room to design. Always sounds like a small box (boominess).
Control room	0.2—0.4 sec.	If the room is also used for recording acoustic instruments the reverberation should be slightly longer.
Recording studios	0.4—0.6 sec.	Rhythmic music.
Living room	0.4—0.5 sec.	
Lecture room	0.6—0.9 sec.	Provides level to the speech but keeps it intelligible.
Cinema (larger)	0.7—1.0 sec.	Must provide a fair reproduction of the audio.
Rock 'n' roll (smaller venues)	0.6—1.6 sec.	Linear relation. Room sizes from 1,000 m^3 to 10,000 m^3 or 35,300 ft^3 to 353,000 ft^3
Theater	1.1—1.4 sec.	Provides level to the speech but keeps it intelligible.
Opera	approx. 1.6 sec.	The reverberation must sustain the singing but still retain some degree of intelligibility.
Concert hall for classical music	1.8—2.2 sec.	May vary with size of hall and music genre.

When designing rooms for audio, especially the smaller ones, the target reverberation time is linked to room size. In commonly used recommendations for listening rooms/control rooms (for instance EBU Tech. 3276 and EBU Tech. 3276 supplement 1), the nominal reverberation time T_m is defined as:

$0.2 < T_m < 0.4$ [s], with

$T_m = 0.25(V/V_0)^{1/3}$ [s]

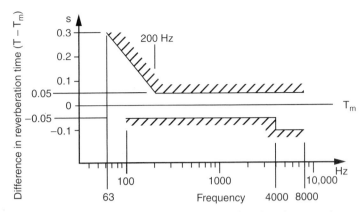

FIGURE 31.5 General recommendation for reverberation time for control rooms with indication of the tolerance limits (Reference EBU Tech 3276).

where

V = the volume in cubic meters (or cubic feet)

V_0 = a reference volume for 100 m^3 (or 3531 ft^3)

Additionally, limits to the variation of the reverberation time vs. frequency is given. In Figure 31.5 these limits are shown.

There are a number of standards and recommendations for listening rooms and control rooms. "IEC 60268-13: Sound system equipment — Part 13: Listening tests on loudspeakers" is one of them. Another one is given by the THX PM3 proprietary standard.

For larger auditoria the relation between the reverberation time at lower and higher frequencies is sometimes expressed as the Bass Ratio, BR, calculated as follows:

$$BR = \frac{T_{125Hz} + T_{250Hz}}{T_{500Hz} + T_{1000Hz}}$$

where

$T_{125\,Hz} \ldots T_{1000\,Hz}$ is the reverberation time in the octave bands 125 Hz ... 1000 Hz, respectively.

For classical music the BR should be in the range of 1.0–1.3. For speech and rhythmic music this value should be in the range of 0.9–1.0.

CALCULATING THE REVERBERATION TIME

When calculating the reverberation time we utilize a simple equation known as the Sabine equation:

$$T = \frac{0.161 \cdot V}{A} [s] \text{ (unit: meter)}$$

or

$$T = \frac{0.049 \cdot V}{A} [s] \text{ (unit: foot)}$$

where

T = the reverberation time in seconds

V = the volume in cubic meters (or cubic feet)

A = the absorption $(\alpha_1 \cdot S_1) + (\alpha_2 \cdot S_2) + (\alpha_3 \cdot S_3) + ... + (\alpha_n \cdot S_n)$

where

α = the sound absorption coefficient of a given part of the surface/material $(0 \leq \alpha \leq 1)$

S = the area of that given surface/material.

A sound absorption coefficient of 1 $(\alpha = 1)$ is the same as an open window by which the sound leaves and never returns. An absorption coefficient of 0 $(\alpha = 0)$ is comparable to that of a hard reflective surface like concrete; sound hitting the surface is reflected by 100%.

The unit for absorption is the sabin; so one square meter of a material that has an absorption coefficient of one equals one square meter sabin. For example:

1 m^2 of a material with $\alpha = 1.00 \Rightarrow 1.00$ m^2 sab.

3 m^2 of a material with $\alpha = 0.50 \Rightarrow 1.50$ m^2 sab.

And similarly in feet:

1 ft^2 of a material with $\alpha = 1.00 \Rightarrow 1.00$ ft^2 sab.

3 ft^2 of a material with $\alpha = 0.50 \Rightarrow 1.50$ ft^2 sab.

This absorption "A" is normally defined in frequency bands, that is, octave bands or 1/3 octave bands.

The use of the simple Sabine equation is an approximation and it assumes that the absorption material is well distributed among the faces of the room in order to get the result right. Still, if we try to calculate the reverberation time of a room that is totally treated with a material of 100% absorption, the reverberation time is *not* 0.0 seconds. If the volume is in the range of 45–50 m^3 (1600–1750 ft^3) the reverberation time calculated is around 0.1 s.

The Sabine equation has been modified in order to reduce uncertainties. Two of the most utilized modifications follow:

Eyring equation:

$$T = \frac{0.161 \cdot V}{-s \cdot \ln(1 - \overline{\alpha})} \; [s] \;\; (\text{unit: meter})$$

$$T = \frac{0.049 \cdot V}{-s \cdot \ln(1 - \overline{\alpha})} \; [s] \;\; (\text{unit: foot})$$

Fitzroy equation:

$$T = 0.161 \cdot \frac{V}{s^2} \left[\frac{-x}{\ln(1 - \alpha_x)} + \frac{-y}{\ln(1 - \alpha_y)} + \frac{-z}{\ln(1 - \alpha_z)} \right] \; [s] \;\; (\text{unit: meter})$$

$$T = 0.049 \cdot \frac{V}{s^2} \left[\frac{-x}{\ln(1 - \alpha_x)} + \frac{-y}{\ln(1 - \alpha_y)} + \frac{-z}{\ln(1 - \alpha_z)} \right] \; [s] \;\; (\text{unit: foot})$$

where

x, y, and z are the three sets of parallel walls in a (box-shaped) room (the end walls, the side walls, and the floor and ceiling).

The standard absorbing coefficients provided for the room designer are normally obtained by placing the materials in a special reverberant chamber. The reverberation time is measured with a known area of the material placed in the room. The material is removed and the reverberation time is measured again. Knowing the volume of the room, the absorption is calculated from the reverberation time difference. That is the standard methodology. However, the characteristics of a given absorbing material should also be measured in a space providing the same conditions as the room to be designed. This is necessary at least for the frequency range below the Schroeder frequency.

ABSORBERS

We use sound absorbing materials when we want to control the acoustics (i.e., the reverberation time of a room). However, we must be aware of the fact that different materials may exhibit absorption in different parts of the frequency range. In general there are three groups of absorbers at hand for the acoustic treatment of the room: porous absorbers, resonance absorbers, and membrane absorbers.

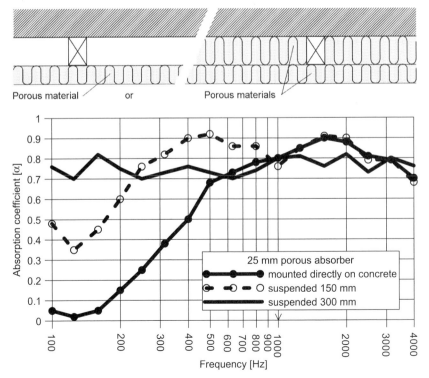

FIGURE 31.6 Porous absorber and generic absorption curve.

Porous Absorbers

These absorbers include materials like foam, cloth, mineral wool, polyester fiber, etc. They absorb frequencies in the high range and are very efficient at doing so with absorption coefficients typically above 0.7. If the porous absorbers are placed at a distance (25−30 cm or approximately 1 ft) from a hard surface (wall or ceiling), however, they will provide efficient absorption down to approximately 100 Hz. The porous absorbers are efficient when the depth is at least one quarter of the wavelength that is to be absorbed. So putting a carpet on the floor of a room will not damp low frequencies.

In some cases the specification sheets of fancy-looking products like foam with a pyramid-shaped surface may exhibit an α-value above 1.00. This is not an error. However, this high absorption exists only above a certain frequency. This is due to the fact that the high frequencies can "see" a larger surface (the pyramid faces) in comparison to the area that the material actually covers. The lower frequencies can only "see" the average depth of the absorber (hence this absorber might be more efficient at lower frequencies if turned inside out).

Resonance Absorbers

Resonance absorbers are based on the principles of the Helmholtz resonator and include perforated plates, perforated bricks, slitted walls, etc. They are efficient

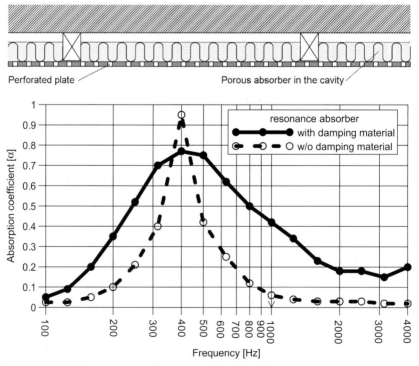

FIGURE 31.7 Resonance absorber and generic absorption curve.

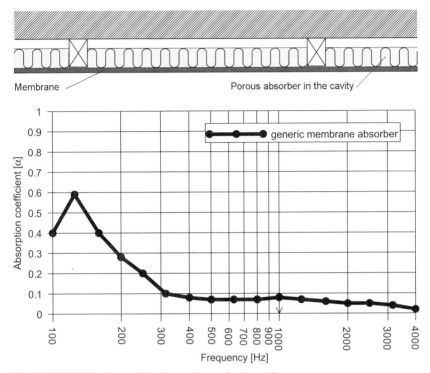

FIGURE 31.8 Membrane absorber and generic absorption curve.

in the midrange (200 Hz − 5 kHz). The advantage of this absorber is the possibility of tuning the resonance. However, normally it is preferred to design the absorber to have broader band absorption around the resonance frequency by adding damping porous materials to the chamber.

Membrane Absorbers

Membrane absorbers work at low frequencies. They can basically be a part of the building construction: light walls, windows, floating floors, etc. They are not very efficient. At the resonance frequency the absorption coefficient is in the range of 0.2−0.3 (measured in a diffuse sound field). However, if the membranes are a part of the construction you already may have large areas available.

In practice, when placing the membrane absorber in a room, the apparent absorption coefficient of an efficient membrane absorber may range from below 0.5 to 2.5, depending on its placement in the room! The membrane absorber is most efficient when placed in the corners where the maxima (pressure) of the room modes are found.

SOUND DISTRIBUTION

Requirements regarding an appropriate sound distribution are normally related to larger rooms, like auditoriums, concert halls, etc. However, in smaller rooms

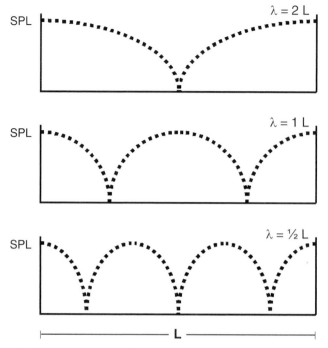

FIGURE 31.9 First-, second-, and third-order room modes in one dimension.

this requirement can be related to frequencies in the modal region, as mentioned earlier. In most rooms for audio it is vital to obtain a good low frequency response not just in one single point, but over a wider listening area or at least in a couple of listening positions.

Figure 31.9 shows a graphic representation of the first three axial modes — also called standing waves — through a single room dimension for an instant in time. Sound pressure maxima always exist at the room boundaries (i.e., the left and right side of Figure 31.9). The second-order mode has a maximum at the center as well, while the first- and third-order modes pass through a minimum at this point. The point where the sound pressure drops to its minimum value is commonly referred to as a "null." If there is no mode damping at all, the sound pressure at the nulls drops to zero. However, in most real rooms the response dips at the nulls are typically in the −20 dB range.

What we find in practice is very often rooms that are not exactly box shaped, and rooms that do not have identical absorption on opposite walls. This may be a result of an attempt to prevent (reduce) the influence from the standing waves. This is done, for instance, by building non-parallel walls. Just a five degree offset of a wall can change the build-up of standing waves dramatically. This is generally fine; however, we can get unpredictable results when there is no longer symmetry along the median plane. Thus, we may end up with different responses for each of the speakers in a stereo or a 5.1 channel setup, making it very difficult to calibrate for proper monitoring. It is worth noting that there always are standing waves (modes) present in a room. It is just a question of how strong the modes are.

The sound distribution in larger rooms like auditoria for speech and music is related to the amount and equality of the sound received and perceived across the audience area or across the stage. Some of the measures are described in the "Large Room Parameters" section.

BACKGROUND NOISE

In general most control rooms and studios have to be kept quiet. Noisy equipment must be built into racks having airtight doors and the cooling system is placed somewhere else. Hopefully!

However, from time to time cables seem to grow out of the rack box so the door cannot be closed properly. In larger studios, concert venues, and churches it is the HVAC or outside traffic noise that set the limits for the activity. (See Chapter 30 on NR curves for noise specifications.)

ECHOES AND FLUTTER ECHOES

Single reflections that arrive 30—50 ms (or later) after the initial sound are perceived as echoes. This is a typical experience when playing music from a stage and receiving the echo from a reflective rear wall of the room. However, a real (slap-) echo will never occur in smaller rooms due to the limited distances between the walls.

Flutter echoes, however, exist in all room sizes but mostly in smaller rooms. They can occur between parallel surfaces like two walls, between control room windows, between floor and ceiling, between control room windows and the backside of loudspeakers, between computer monitors, between table and lamp screens, etc. The repeated reflections create a kind of tone themselves.

Flutter echoes should never be present in any room for audio. The phenomenon is easy to remove: offset parallel surfaces; supply sound absorption if necessary; or install diffusing elements that remove the flutter but retain the sound energy in the room.

EARLY REFLECTIONS

The definition of early reflections and their influence on the perceived sound depend on the size of the room and thus the level and time of arrival. In a small room early reflections in general should be avoided. In a large room (for music) early reflections provide valuable information on room size and directional definition.

Early Reflections in a Small Room

Early reflections in a small room may affect the perceived frequency response in the listening position, typically in the midrange. The problem is comb filtering and it should always be avoided. (See Chapter 22 on summation of audio signals.)

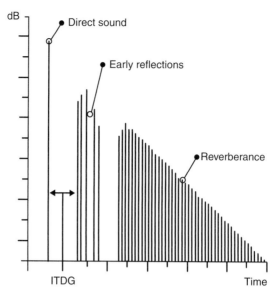

FIGURE 31.10 A reflectogram showing the direct sound, the initial time delay gap (ITDG), the early reflections, and the reverberance tail.

It is not always the best solution to remove reflections by adding absorption. In most rooms too much porous absorption is introduced. The result is a room with too short a reverberation time at higher frequencies. It is a better idea to involve diffusion in order to retain the high frequency energy in the room. There are several solutions for this, the Schroeder diffuser being one. These devices may introduce diffusion in one or two dimensions. Most diffusers also exhibit some absorption in the active range.

Early Reflections in the Large Room

In a large room, especially a concert hall, early reflections are important both to the listeners in the room as well as to the musicians and even the recording engineer when placing and aligning microphones.

At a given distance from a sound source the first sound received is the direct sound. This is followed by individual early reflections; later the higher order reflections arrive from all directions and are characterized as the diffuse sound field. The time span between the direct sound and the early reflections is called initial time delay gap or ITDG.

LARGE ROOM PARAMETERS

In auditoria design and evaluation there has always been an effort to find quantities that objectively describe parameters that express what is subjectively perceived. Some of these measures are briefly mentioned here and in general describe the relation between a sound source — it can either be a generic sound source, a human voice, a musical instrument, a loudspeaker, etc. – and the sound field received in the point of observation. Several of these parameters can be calculated in simulation software like EASE® and ODEON® that are used by many sound engineers.

D/R ratio

The D/R ratio expresses the ratio of the level of direct sound from a sound source received in a given position of the room to that of the level of reflected/diffuse sound received in the same position. The directivity of the sound source has to be considered.

Reverberation Radius and Critical Distance

The reverberation radius r_H indicates the positions in the room where the amount of direct sound equals the amount of reflected sound (D/R ratio $= 1$). For a spherical sound source, the reverberation radius r_H can be expressed by the following:

$$r_H = \sqrt{\frac{A}{16\pi}} \approx \sqrt{\frac{A}{50}} \approx 0.141\sqrt{A} \approx 0.057\sqrt{\frac{V}{T}}$$

where
A = the absorption in m^2 or ft^2
V = the volume in m^3 or ft^3
T = the reverberation time in seconds.

The name reverberation "radius" (in German Hall Radius, hence r_H) is only valid if the positions around a given sound source form a fixed radius. However, this is not the case if the sound source is directive (as is for instance with PA/SR speakers). Hence we use the expression critical distance, r_R or D_c:

$$D_z \approx 0.141\sqrt{A - Q_s} \approx 0.057\sqrt{\frac{V \cdot Q_s}{T}}$$

where
Q_s = the directivity factor of the sound source
An omnidirectional source has a Q_s of 1.

The critical distance from a (monitor-) loudspeaker will change with frequency as most speakers are more or less omnidirectional at low frequencies but directive at higher frequencies.

Clarity, C$_{80}$, C$_{50}$, C$_7$

Clarity is a way to evaluate and compare the amount of first arriving sound to later arriving sound. Originally this was related to predicting the clarity of different modes of music and expressed as the ratio in dB of the energy arriving before and after 80 ms relative to the time of first arrival: C_{80}

Later other measures using the same methodology were introduced:

C_{50}: The ratio in dB of the energy arriving before and after 50 ms relative to the first sound. This calculation is used to predict the articulation of human speech.

C_7: The ratio in dB of the energy arriving before and after 7 ms relative to the first sound. This calculation is used to predict the strength of direct sound sources for localization in auditoria.

Strength, G

The sound strength (or relative sound level) G is measured using a calibrated omnidirectional sound source and is the ratio in dB of the sound energy of the measured impulse response to that of the response measured at a distance of 10 m from the same sound source in a free field.

The strength measurement is a practical way to evaluate the sound distribution in an auditorium. For further analysis the received sound can be evaluated in relation to time of arrival. For example, G_{80} expresses the strength of the sound within the first 80 ms relative to the arrival of the first sound. The sound strength of the late-arriving sound, G_L, consists of sound energy arriving at the receiver more than 80 ms after the direct sound.

Further analyses of early and late arrival of lateral sounds, G_{EL} and G_{LL}, respectively, can be obtained by exchanging the standard omnidirectional measurement microphone with a figure-of-eight microphone. Performing measurements perpendicular to the main axis can provide information on the envelopment of the sound.

Bibliography

Bech, S. Perception of reproduced sound: Audibility of individual reflections in a complete sound field, II. AES 99th Conv. 1995. Preprint 4093.

Beranek, Leo L. Sabine and Eyring equations. JASA, vol. 120, No. 3. 2006.

Cox, T.J. and D'Antonio, P. Acoustic Absorbers and Diffusers, Spon Press, London 2004.

EBU Tech. 3276 and EBU Tech. 3276 supplement-1: Listening conditions for the assessment of sound programme material: multi channel sound.

Fuchs, H.V. Zur absorption tiefer Frequenzen in Tonstudios. (About absorption at lower frequencies in recording studios.) Reprint from Runfunktechnische Mitteilungen. Nummer 36, Heft 1, pp. 1−10. 1992.

Geddes, Earl R. Small Room Acoustics in the Statistical Region. Proceedings of the AES 15th International Conference: Audio, Acoustics & Small Spaces. 1998.

Hunecke, J., Fuchs, H.V., Zhou, X., Zhangg, T. Einsatz von Membran-Absorbern in der Raumakustik. (Properties of membrane absorbers in room acoustics.) Proceedings of 17th Tonmeistertagung, 1992.

IEC 60268-13 Sound system equipment − Part 13: Listening tests on loudspeakers.

ISO 3382−1:2009; Acoustics − Measurement of room acoustic parameters − Part 1: Performance spaces.

Manual for EASE 4.3 (Enhanced Acoustic Simulator for Engineers) by Ahnert Acoustic Design/AFMG Technologies GmbH. 2008.

Randeberg, R.T. Adjustable Slittet Panel Absorber. Acta Acoustica united with Acoustica. Vol. 88, pp. 507−512. 2002.

Rasmussen, B., Rindel, J.H., and Henriksen, H. Design and Measurement of Short Reverberation Times at Low Frequencies in Talk Studios. J. Audio Eng Soc., vol. 39, issue 1-2 Jan./Feb. (1991).

Sabine, W. C. Collected Papers on Acoustics. Dover Publications, Inc. New York 1964.

Schroeder, M. R. Frequency-Correlation Functions of Frequency Responses in Rooms. JASA, vol. 34, no. 12 1962.

Schroeder, M. R. Diffuse Sound Reflection by Maximum-length Sequences. JASA, vol. 57. January 1975.

Voetmann, J., Klinkby, J. Review of the Low-Frequency Absorber and Its Application to Small Room Acoustics. AES 94th Conv. 1995. Preprint 3578.

Voetmann, J. 50 Years of Sound Control Room Design. AES 122th Conv. 2007. Preprint 7140.

Walker, R. Optimum Dimension Ratios for Small Rooms. AES 96th Conv. 1994. Preprint 4191.

Walker, R. Room Modes and Low Frequencies Responses in Small Enclosures. AES 100th Conv. 1996. Preprint 4194.

Walker, R. A Controlled-reflection Listening Room for Multi-channel Sound. AES 104th Conv, 1998. Preprint 4645.

Glossary

3D sound See *surround sound*.

5.1 surround sound An audio format involving five channels of full bandwidth audio: center, left, right, left surround and right surround. A special low frequency effects channel (LFE or ".1") covers a frequency range from 20 Hz to 120 Hz. The basic setup for music production is defined in ITU 775.

A-D Analog-Digital.

AAC Advanced Audio Coding. Special variant of the MPEG standard.

AB (1) Microphone placement for spaced microphones intended for time difference stereo. (2) Manner in which microphones are provided with power. Also called Tonader. (3) A method for comparison of two versions.

absorption A property of materials that reduces the amount of sound energy reflected; unit: sabin. 1 m^2 full absorption (like an open window) equals 1 m^2 sabin.

absorption coefficient (symbol: α) The practical unit between 0 (no absorption) and 1 (full absorption) expressing the absorbing properties of a material. This is basically specified per octave or 1/3 octave. Absorption may exceed 100% (or $\alpha > 1$) when the surface area seems larger than the area it covers (applies typically only at high frequencies).

AC Alternating current, as opposed to: DC, direct current.

AC-3 Digital Audio Compression Standard.

acoustics The interdisciplinary science that deals with the study of sound.

ADC Analog to Digital Converter. A circuit that converts an analog signal to a digital signal.

ADPCM Adaptive Differential Pulse Code Modulation. Encoding form for digital signals.

ADR Automatic Dialog Replacement. Automatic replacement of location dialog with studio dialog.

AES Audio Engineering Society. International society for professional audio engineers.

AES/EBU The AES (Audio Engineering Society) and EBU (European Broadcast Union) created jointly a standard for the transfer of two channels of digital sound. The signal is bi-phase modulated, self-clocking and runs on a balanced cable (max. 50 m or 150 ft) or via optical fiber. The standard currently is called AES3.

AF Audio Frequency. Audible frequencies, basically in the range of 20 Hz to 20,000 Hz.

A-filter See *A-weighting*.

AFM Designation for a video recorder's frequency-modulated sound track (Audio-FM).

A-format A microphone arrangement with four coincident cardioid capsules. Through use of a dedicated matrix this signal can be converted to B-format. See *sound field microphone*.

AGC Automatic Gain Control. Method by which the amplification in a circuit is controlled by the input voltage or another parameter.

AIFF File format for sound.

AM Amplitude Modulation. Transfer of information by the variation of the amplitude of a carrier wave.

ambience Spatial effect, notably mixing of remotely placed microphones in order to include the atmosphere of the room on the recording.

ambient noise (Acoustic) noise from the surroundings.

Ambisonics A principle that utilizes the sound field microphone involving B-format recording (X, Y, Z, and W).

ampere, A Unit for electrical current.

amplitude For an alternating signal; the size of the variation.

analog Quantities in two separate physical systems having consistently similar relationships to each other are called analogous. One is then called the analog of the other (for example, sound pressure in front of a microphone and the electrical output of that microphone).

anchor element The perceptual loudness reference point or element around which other elements are balanced in producing the final mix of the content, or that a reasonable viewer would focus on when setting the volume control.

anechoic room Room without echo; the reverberation time is ideally close to zero seconds.

ANSI American National Standards Institute. A federation of American organizations concerned with the development of standards.

anti-aliasing Low-pass filter for the removal of frequencies that otherwise would create "alias" frequencies that were not present in the signal originally.

apt-X Bit reduction system from Audio Processing Technologies used by DTS. 4:1 or 3:1 compression.

artifact Unwanted effects that arise due to technical limitations.

ASCII American Standard Code for Information Interchange. Character set encoding used in data transfer.

asynchronous sample rate converter When two (digital) devices cannot be synchronized (for example, a CD player with a mixer) even a small deviation between the clock frequencies of the two devices will cause occasional glitches due to the accumulation of shortages or excesses. This creates a small "crack" in the sound. The asynchronous sample rate converter can perform interpolation, and hence create the missing intermediate values so that glitches are avoided.

ATR Audio Tape Recorder.

ATRAC Adaptive Transform Acoustic Coding. Bit reduction system from Sony for use on MiniDisc. In stereo, 300 kbps is used. The system is also used for films (see *SDDS*).

ATSC Advanced Television Systems Committee.

attack The beginning of a sound; the initial transient of a musical note.

attenuation Variable or fixed downward adjustment of a signal level.

attenuator A device used to control the level of an electrical signal.

audio That part of sound technology that concerns itself with the recording and reproduction of sound.

autolocate Facility on a tape unit's transport system.

AUX Auxiliary jack, an extra input or output.

A-weighting The connection of an A-filter (IEC-A) for the acoustical and electrical measurement of noise. Produces a result that roughly corresponds to the human ear.

azimuth Angle of a magnetic head gap in relation to the direction of travel of a magnetic tape.

balancing When a signal is run independently of the frame/shielding and both terminals have the same impedance to the frame. See *symmetry attenuation*.

band-pass filter See *filter, band-pass*.

bandwidth The distance between the 3 dB cutoff frequencies on a response curve. Expressed either in octaves or in Hz.

batch The number of sound or video media produced in the same fabrication process.

BCD Binary Coded Decimal. Decimal digits converted to the base two system.

bel, B Relative logarithmic unit for the measurement of sound levels. Normally one-tenth of a bel is regarded as the main unit: decibel, dB.

Betacam Video format that uses the Betamax cassette with a tape speed of 10.15 cm/s [4 in./s]. and two analog sound tracks. Betacam SP: two analog and two FM modulated sound tracks. Betacam SP Digital: one analog, two FM, and two digital sound tracks.

B-format (microphones) A recording format virtually based on three orthogonal oriented figure 8 microphones (**X, Y, Z**) and one omnidirectional microphone (**W**). By combining these signals a variety of directional characteristics can be obtained. Most B-format microphones are designed with cardioid microphones. See *sound field microphone*.

bias High-frequency alternating current (typically 100 kHz) that is added to the recording head in a tape recorder together with the sound signal. The use of bias improves the quality of the recording with respect to both the frequency response and the distortion.

binary Number system that only contains two digits.

binaural recording Stereo recording using either an artificial head with built-in microphones or a real person with small microphones in, or close to, the ear canals. For correct monitoring, the recording must be reproduced by using headphones to eliminate crosstalk.

biphase modulation A form of modulation that, among other things, is used by the AES/EBU interface and time code. In the bit stream, each bit shift is marked with a level shift (from high to low or vice versa). If the bit value is high, this is marked with a level shift in the middle of the bit concerned. Some of the advantages include that the DC component in the signal is minimized and the signal is self-clocking.

bit Binary digit in digital technology, "0" or "1".

bit companding A technique for digital audio via which greater contrast can be obtained for a given number of bits.

bit reduction As linear quantization can result in a larger number of bits per second than there is room for in a transmission channel or on a storage medium, the number of bits is reduced, preferably in a manner so that it cannot be heard in the audio signal.

bit Smallest unit in digital technology, with a value of either 0 or 1.

bit stream converter See *Delta-Sigma converter.*

BNC connector Baby N-Connector; coaxial connector.

BS.1770 Formally ITU-R BS.1770. This specifies an algorithm that provides a numerical value indicative of the perceived loudness of the content that is measured. Loudness meters and measurement tools that have implemented the BS.1770 algorithm will report loudness in units of "LKFS."

buffer Circuit for the maintenance or improvement of a function.

bus A common set of conductors where many signals are gathered, for example in a mixer console.

byte Data word, consisting of a number of bits, normally 8.

c/s Cycles per second = Hz.

calibration The process of measuring to determine the accuracy of the measurement chain.

calibrator, acoustic A device that produces a known sound pressure on a microphone in a sound level measurement system.

camcorder A contraction of the words camera and recorder; the recording medium and the camera are built together as a unit.

cartridge Special tape cassette with 1/4-inch tape earlier used for jingles, ads, etc., at radio stations. Format: mono, stereo, or 8-track versions.

CC See *Compact Cassette.*

center frequency The arithmetic center of a constant bandwidth filter, or the geometric center (midpoint on a logarithmic scale) of a constant percentage bandwidth filter.

center post The center lead in a contact device.

channel separation The attenuation of one channel appearing in neighboring channels.

clean sound Atmospheric sound.

clipper See *limiter.*

clipping An electrical signal is clipped if the signal level exceeds the capabilities of the signal chain or recording device. It is a distortion of the signal.

comb filter A distortion produced by combining an electrical or acoustic signal with a delayed replica of itself. The result is a series of tops and dips across the frequency response that makes it look like a comb.

Compact Cassette , CC. Registered name for the cassette developed by Philips with a 3.81 mm [0.15 in.] audio tape.

compand Describes a process that involves compression and expansion successively.

compander A contraction of compressor and expander. A device that can perform both functions.

compression Reduction of the dynamic range of recorded audio.

compressor A device or plug-in that provides reduction of the dynamic range of recorded audio.

content 0 Material or essence used for distribution by an operator.

corner frequency Transition frequency of a filter.

CRC Cyclic redundancy check. Error correction.

crest factor The term used to represent the ratio of the peak (crest) value to the RMS value of a waveform. For example, a sine wave has a crest factor of 1.4 (or 3 dB), since the peak value equals 1.414 times the RMS value. Music has a wide crest factor range of 4–10 (or 12–20 dB). This means that music peaks can be 12–20 dB higher than the RMS value, which is why headroom is so important in recording and in audio design.

critical band In human hearing, only those frequency components within a narrow band, called the critical band, will mask a given tone. Critical bandwidth varies with frequency but is usually between 1/6- and 1/3-octave. The ears act like a set of parallel filters, each with its own bandwidth.

critical distance, D_c The distance from a sound source in a room at which the direct sound and the diffuse, reflected sound has the same level.

cross fade The audio equivalent of what is known as dissolve in video.

crossover frequency In a loudspeaker with multiple radiators, the crossover frequency is the −3 dB point of the network dividing the signal energy.

crosstalk When a recording on a different track can be heard on a track.

cue Audio or visual information that concerns timing or synchronization.

cue wheel Control button for use with slow forwards and backwards winding with cueing.

cutoff The cutoff frequency of a filter. The frequency at which a filter begins to attenuate. Often defined by the attenuation being 3 dB at the frequency concerned.

D-A Digital-Analog (digital to analog).

DAC Digital-to-analog converter. A circuit that converts a digital signal to an analog signal. The abbreviation D-A is often used.

DAT Digital Audio Tape (-recorder). Digital tape format (cassette). Normally understood as R-DAT (Rotary head DAT) as opposed to S-DAT (Stationary head DAT), which never became a widespread format.

dB See *decibel*.

dB FS Decibels, relative to full-scale sine wave (per AES17).

dB TP Decibels, true-peak relative to full scale (per ITU-R BS.1770 Annex 2).

dBm Logarithmic relation with a reference of 1 mW/600 ohm.

dBu Logarithmic relation with a reference of 0.775 V.

dBV Logarithmic relation with a reference of 1 V.

dbx A particular brand of audio processing equipment including noise reduction.

DC Direct current, as opposed to AC, alternating current.

DC offset The change in input voltage required to produce zero output voltage when no signal is applied to a device. Basically this is an unwanted phenomenon in audio; however, DC offset often occurs in lower quality sound cards.

D-connector Multi-conductor plug with D-shaped collar.

decade Ten times any quantity or frequency range. The range of the human ear is about 3 decades.

decay rate A measure of the decay of acoustic signals, expressed as a slope in dB/second. Essentially, the rate at which a signal drops off.

decibel, dB Logarithmic level specification.

decimation Suitable restructuring of data, for example swapping bandwidth (sampling frequency) for bit depth.

de-emphasis See *emphasis*.

de-esser Signal processing to reduce "S"es, typically in vocal recordings.

delay Time delay. (1) Electrical circuit that can delay a signal, in practice from fractions of a millisecond up to multiple seconds; used for sound effects. (2) Unwanted effect of time-consuming digital conversion, processing, etc.; see also *latency*.

Delta-Sigma converter ($\Delta\Sigma$ converter) Serial conversion at a high sampling frequency, where each bit specifies whether the current value of the signal is higher or lower than the prior one. After this process, this bit stream is converted to standard values, for example 16-bit format.

DI (1) Digital In, digital input. (2) Direct Injection, intermediate amplifier. (3) Directivity Index, directional index for acoustic transducers.

dialnorm An AC-3 metadata parameter, numerically equal to the absolute value of the dialog level carried in the AC-3 bit stream.

dialog level The loudness, in LKFS units, of the anchor element.

diffraction The distortion of a wave front caused by the presence of an obstacle in the sound field; the scattering of radiation at an object smaller than one wavelength and the subsequent interference of the scattered wave fronts.

diffuse sound The sound field contains no directional information and spreads randomly in the room.

diffuser A device that provides scattering when sound hits the surface.

digital Concerning a state wherein an electrical signal has been converted to a series of impulses according to a specific code. The signal has come to exist in "tabular form."

digital interface Interface between digital systems. A number of standards exist, of which SP-DIF and AES/EBU are the most used.

DIN plug Contact connection as per the DIN norm.

DIO Digital In/Out, digital input and output.

Directivity factor The ratio of the mean-square pressure (or intensity) on the axis of a transducer at a certain distance to the mean-square pressure (or intensity) that a spherical source radiating the same power would produce at that point.

Directivity Index (DI) Directivity factor expressed in dB ($10 \cdot \log$ (directivity factor)).

dissolve Mixing (of images). Corresponds to crossfade in audio.

distance double law When doubling the distance from a point source, the sound pressure is halved (reduced by 6 dB).

distortion When a signal is changed from its original form, for example due to nonlinearities in the transmission chain.

distortion factor Percentage measure for harmonic distortion, for example, $k3 = 3\%$ means that the third harmonic overtone is 3% of the fundamental tone.

distribution amplifier Amplifier that can distribute a signal to multiple inputs without the signal source becoming overloaded.

dither Noise that is added to the lowest bit in the digital signal in order to reduce the distortion at low levels of the audio signal. Dither can be noise shaped, in order to be less audible.

DO Digital Out, digital output.

Dolby E An audio data-rate reduction technology designed for use in contribution and distribution that also conveys Dolby E metadata.

Doppler effect or Doppler shift The apparent upward shift in frequency of a sound as a noise source approaches the listener or the apparent downward shift when the noise source recedes; e.g., when a speedy car passes by.

DRC Dynamic range control.

DRC profile A collection of parameters that describe how dynamic range control metadata is calculated.

drop frame Variant of time code, where a frame is periodically skipped in order to preserve synchronization.

drop-out Short-duration loss of the signal on a tape due to faults in the tape's magnetic coating.

DSB Digital satellite broadcasting.

DSP Digital signal processor. A circuit that can perform manipulation of data.

D-sub A plug type that can be used, among other things, for the transfer of digital signals. The flange around the connector pins are D-shaped, hence the name.

DTS Digital Theater System. Digital sound system originally for movies. The sound information is stored on a CD-ROM, which is controlled by a time code on the film. Uses APT-X bit reduction.

dubbing Mix of one signal with another.

ducking Automatic compression, for example when a speaker's insert dims the music signal.

dynamic range The relationship between the strongest and weakest passages in the program material. Used for both the acoustic and the electrical signal.

EBU European Broadcasting Union. An association of European radio broadcasting stations.

echo Sound impulse arising from a reflection with such a strength and time delay after the direct sound that it is perceived as a repetition of it.

edge track Designation for the longitudinal (sound) tracks of a video tape, since these are located on the edge of the tape.

EFM Eight-to-fourteen modulation. Digital modulation form, used for CDs among other things.

EFP Electronic field production. The production form typically applied for drama and documentaries produced outside the studio.

EIAJ Electronic Industry Association of Japan. An interest organization in Japan that issues some norms.

emphasis A technique often used in analog transmission systems (wireless microphones, FM broadcast, etc.) to enhance dynamic range by raising the level of higher frequencies. De-emphasis is introduced on the receiving side.

ENG Electronic news gathering. News production in which electronic media is used, i.e. video tape, HD, or flash recorders (as opposed to film).

EQ See *equalizer, graphic; equalizer, parametric; equalizing.*

equal loudness contour A contour representing a constant loudness for all audible frequencies. The contour with a sound pressure level of 40 dB at 1000 Hz is arbitrarily defined as the 40-phon contour.

equalizer, graphic Electronic equipment for "equalizing." Built from a number of 1/1 octave or 1/3 octave band-pass filters that can each amplify or attenuate, and used to obtain a desired frequency response.

equalizer, parametric. Electronic equipment for "equalizing". Constructed from a set of filters where the center frequency, bandwidth and amplification/attenuation can be adjusted independently for each filter.

equalizing (1) The process that consists of modifying the frequency balance in the amplifier chain for the purpose of obtaining a flat frequency response, minimizing noise, or achieving an artistic effect. (2) Equalization of nonlinearity (in frequency response).

expander (1) Electronic equipment in which the output signal's dynamic range is increased in relation to that of the input signal. (2) Designation for a controllable synthesizer without a keyboard.

far field A region in free space at a much greater distance from a sound source than the linear dimensions of the source itself where the sound pressure decreases according to the inverse-square law (the sound pressure level decreases 6 dB with each doubling of distance from the source).

FFT Fast Fourier Transform, an efficient method of estimating the frequency spectrum of a signal.

file-based scaling device A device used to apply an overall gain correction to audio content stored as files.

filter A device or algorithm for separating components of a signal on the basis of their frequency. It allows components in one or more frequency bands to pass relatively unattenuated, and it attenuates components in other frequency bands.

filter, band-pass A filter that passes all frequencies between a low-frequency cutoff point and a high-frequency cutoff point.

filter, high-boost A filter that amplifies frequencies above a specific frequency.

filter, high-cut A filter that attenuates frequencies above a specific cutoff frequency.

filter, high-pass A filter that passes all frequencies above a cutoff frequency but attenuates low-frequency components. They are used in instrumentation to eliminate low-frequency noise, and to separate alternating components from direct (DC) components in a signal.

filter, low-cut A filter that cuts off low frequency signals below the cutoff frequency with a certain attenuation (roll off). In microphones the filters are typically active below 80−300 Hz, and the slope is typically 6 or 12 dB/octave. See also *filter, high-pass.*

filter, low-pass A filter that passes signals below the cutoff frequency and attenuates the signal above that frequency. An antialiasing filter in a digital system is an example of a low-pass filter with a very steep roll off.

filter, notch Narrow-band filter with very strong attenuation in a very narrow frequency range. Used to remove individual frequencies, for example hum.

filter, octave Filter with a bandwidth of an octave.

filter, shelving A type of filter that gives constant amplification or attenuation from the corner frequency.

filter, third-octave A filter whose upper-to-lower pass band limits bear a ratio of $2^{1/3}$, which corresponds to 23% of the center frequency.

flanging Sound effect based upon the direct signal mixed together with itself using varying time delays. Originally made using a tape recorder, where the source reel is slowed down by a finger placed on the flange of the reel.

FM Frequency modulation. Modulation principle in which a carrier wave is varied about its center frequency in proportion to the frequency of the modulating wave and where the oscillation of the carrier wave is proportional to the amplitude of the modulating wave.

fold back The musicians on the scene have a need to be able to hear themselves and the others in a quite specific manner. They are given a fold back or monitor loudspeaker, where the sound is specially mixed for the purpose.

frame In video, an image. The smallest unit of a time code.

framesync Short for "frame synchronizer."

free field An environment in which there are no reflective surfaces within the frequency region of interest and the sound is isotropic and homogeneous.

frequency The number of cycles per second. Its reciprocal is the period. Specified in hertz (Hz).

frequency response Figure that shows the relationship between amplitude and frequency.

frequency weighting Modification of the frequency spectrum of a signal by means of a filter having a conventional characteristic known as A, B, C, D, RLB, K, CCIR/ITU, etc.

fundamental (1) The basic pitch of a musical note. (2) Fundamental frequency, the lowest frequency of a vibrating system. The spectrum of a periodic signal will consist of a fundamental component at the reciprocal of the period and possibly a series of harmonics of this frequency.

gain Amplification (in a circuit).

gate See *noise gate.*

gear Encompasses all equipment.

glitch When a bit is skipped, it can lead to a little "crack" in the sound.

GPI General Purpose Interface.

graphic equalizer See *equalizer, graphic.*

Haas effect Also called the precedence effect or principle of first arrival.

harmonic A discrete sinusoidal (pure-tone) component whose frequency is an integer multiple of the fundamental frequency of the wave. If a component has a frequency twice that of the fundamental, it is called the second harmonic, etc.

harmonic distortion Changing the harmonic content of a signal by passing it through a nonlinear device. Clipping results in harmonic distortion (uneven harmonics).

harmonic overtones Tones with the frequency of an integer multiple of the fundamental frequency.

headroom Overloading reserve. The amount of signal above nominal level that can be permitted before overloading arises with distortion as a consequence.

hertz, Hz Measurement unit for frequency.

HF High frequency.

high-boost filter See *filter, high-boost*.

high-cut filter See *filter, high-cut*.

high-pass filter See *filter, high-pass*.

house sync Sync signal that is distributed (in-house) so all digital devices can run at the same speed.

HVAC noise Heating, ventilation, and air conditioning noise. The word is used in connection with requirements concerning noise in control rooms, cinemas, etc.

Hz See *hertz*.

iCheck Integrity check. Reveals if the signal is spatially compromised, e.g., because of data reduction, such as MP3 or AAC encoded at too low a bit rate.

IEC International Electrotechnical Commission. A standardization commission.

IM Intermodulation.

impedance Electric resistance: a material's opposition to the flow of electric current; measured in ohms.

impedance matching Maximum power is transferred from one circuit to another when the output impedance of the one is matched to the input impedance of the other. Impedance matching is generally only relevant for RF and electrically coupled digital interfaces.

impulse response The response of a system to a unit impulse. The Fourier transform of the impulse response is the frequency response.

induction The property that an electrical current is produced in a conductor when it is exposed to a varying magnetic field.

inductor Coil wound on an iron core.

infrasound Sound at frequencies below the audible range, that is, below about $16 - 20$ Hz.

initial time-delay gap (ITDG) The time gap between the arrival of the direct sound and the first sound reflected from the surfaces of the room.

in-phase Two periodic waves reaching peaks and going through zero at the same instant are said to be "in phase."

insert (1) Insertion of a clip (scene) in an existing recording. (2) A breakpoint in a mixer channel where an external device can be inserted.

interpolation Computation of intermediate values in relation to fixed values (for example, values between two samples).

inverse square law A description of the acoustic wave behavior in which the mean-square pressure varies inversely with the square of the distance from the source. This behavior occurs in free-field situations, where the sound pressure level decreases 6 dB with each doubling of distance from the source.

ips Inches per second.

IR Infrared. IR light is used as a control signal in remote control units.

ISO International Organization for Standardization.

ITU International Telecommunication Union.

jack plug One-legged connector with two or three contacts. Normal dimensions are 6/3.5/2.5 mm.

jack-bay A patch panel with jacks that are used to make connections between different devices easy to establish.

jam sync Method for recording new time codes from a source tape.

jitter Expresses that the sample timing deviates from a fixed rate. Jitter is measured in seconds (typically ns or ps). Jitter causes noise in the signal reproduced. The worst kind of jitter is sampling jitter as this cannot be corrected later. Transmission jitter can to some degree be compensated for.

kbps or kb/s Kilobits per second.

kHz, kiloherz One thousand hertz, see *hertz*.

latency Delay due to processing.

layback A post-production technique where audio is rejoined with the associated video after editing, mixing, or "sweetening."

LCD Liquid crystal display.

LED Light-emitting diode.

Leq Equivalent continuous sound pressure level.

LF Low frequency, (1) Oscillations in the audible frequency spectrum (below 20 kHz). (2) Oscillations in the low frequency part of the audible frequency spectrum.

limiter A signal processing unit that attenuates the output when the input exceeds a given threshold. This provides a fixed maximum output level.

linear phase response The phase is constant with frequency through the device or circuit.

linear quantization Conversion to a digital signal where all bits represent identical steps in level. (As opposed to nonlinear quantization and bit compression).

line driver Intermediate amplifier for studio and stage use that brings a low voltage (for example, from a guitar pickup) up to line level. At the same time, provides a matching for long cables.

line level The voltage that the signals are amplified up to (for example, in the mixer), before they are routed to an output amplifier (or transmitter).

lip-sync The technique of miming to a prerecorded program in connection with a TV recording, etc.

LKFS Loudness, K-weighted, relative to full scale, measured with equipment that conforms to ITU recommendation 1770.

load If an electric circuit has a well-defined output terminal, the circuit connected to this terminal (or its input impedence) is the load.

locate To find a specific position, for example on a track, digital or analog.

long-form content Show or program related material or essence. The typical duration is greater than approximately two to three minutes.

longitudinal Lengthwise.

loudness A perceptual quantity; the magnitude of the physiological effect produced when a sound stimulates the ear.

loudness level (1) In acoustics; measured in phons it is numerically equal to the median sound pressure level (dB) of a free progressive 1000 Hz wave presented to listeners facing the source, which in a number of trials is judged by the listeners to be equally loud. Loudness level can be calculated according to ISO 532B. (2) In audio production; measured in loudness units (LU) according to ITU 1770.

low-cut filter See *filter, low-cut.*

low-pass filter See *filter, low-pass.*

LTC Longitudinal Time Code. The time code that is recorded on a tape's longitudinal track.

LU Loudness Unit. A weighted measure used in program metering. Defined in ITU-1770.

MADI Multichannel Audio Digital Interface; a.k.a. AES10. Interface standard where up to 64 audio channels can be transferred serially on a cable, coax, or optical. NRZI encoding, 125 Mbps.

magnitude The size of a signal.

matrix A circuit in which the output signal is an encoded version of the input signal. A MS-matrix will transform MS to LR or vice versa.

Mbps or Mb/s Megabits per second.

measured loudness The magnitude of an audio signal when measured with equipment that implements the algorithm specified by ITU-R BS.1770/EBU R-128. It is an approximation of perceived loudness.

meter Designation for a measurement instrument for the control of signal levels.

MIDI Music Instrument Digital Interface.

mixing level Indication of the absolute sound pressure level calibration of the mixing studio that produced the content.

mode (1) A room resonance. Axial modes in rectangular rooms are associated with pairs of parallel walls. Tangential modes involve four room surfaces and oblique modes all six surfaces. Their effect is greatest at low frequencies and for small rooms. (2) A specific setting for a device (like record mode, sleep mode, etc.)

modulation An analog process by which the characteristic properties of a wave (the modulating wave) are mixed into another wave (the carrier wave).

modulation noise Noise in the signal chain that varies with the signal strength. Particularly a problem with audio tape recorders.

MOL Maximum output level. For tapes, the highest attainable output.

monitor A control or monitoring device; in this context, a loudspeaker for the assessment of recording quality.

mono Signal in a single channel.

mono compatible A stereo signal, if no significant loss of level and no coloration (comb filtering) occur when summing the channels to one mono signal.

MPEG Motion Picture Experts Group. Standardization organization that sets standards for techniques such as digital audio compression.

MPX Multiplex filter, for removing an FM radio's stereo pilot tone.

MSB Most significant bit.

MS-technique A stereo recording technique using M (mid) and S (side) signals generally from one cardioid and one figure-8 microphone. To obtain L/R format the signals are processed like this: $L = M + S, R = M - S$.

MTC MIDI time code.

MUSICAM Masking pattern adapted Universal Subband Integrated Coding And Multiplexing. An (older) bit reduction system.

mV Millivolt, 1/1000 V.

MVPD Multichannel Video Programming Distributor. Includes DBS service operators, local cable system operators, and cable multiple system operators.

N See *newton*.

NAB or NARTB National Association of Radio and Television Broadcasters. American standardization organization.

NBC Non-backward compatible. Compression standards that are not backwards compatible, i.e., it is not possible to "unpack" stereo to 5.1 channels having first gone from 5.1 to two-channel stereo.

newton Measurement unit for force.

noise (1) Unwanted sound. (2) Technically, sound without tonal components.

noise floor A measure of the signal created from the sum of all noise sources and unwanted signals within an audio system.

noise gate Special signal processing unit whose output signal is 0 until the input signal exceeds a specific, preset value.

noise shaping A special technique providing frequency weighting of dither used in connection with reducing high-resolution digital audio to a lower bit format.

notch filter See *filter, notch*.

NR Noise reduction.

NRZ Non-return to zero.

nWb Nanoweber = 10^{-9} weber. See *weber*.

octave A range of frequencies whose upper frequency limit is twice that of its lower frequency limit. For example, the 1000 hertz octave band contains noise energy at all frequencies from 707 to 1414 hertz. In acoustic measurements, sound pressure level is often measured in octave bands, and the center frequencies of these bands are defined by ISO and ANSI. The sound pressure level of sound that has been passed through an octave band pass filter is termed the octave band sound pressure level. Fractional portions of an octave band are also often used, such as a 1/3 octave band.

octave filter See *filter, octave,*

ohm [Ω] Unit for electrical resistance.

one-bit converter See *Delta-Sigma converter.*

operator A broadcast network, broadcast station, DBS service, local cable system, or cable multiple system operator (MSO).

oscillator Signal generator for the generation of, for example, pure sinusoidal tones with constant amplitude for use in measurements.

output impedance The alternating current resistance a circuit's output represents. May also be called source impedance.

oversampling The use of a sampling frequency that is a number of times higher than is necessary. This makes it easier to make low-pass filters that ensure that no alias frequencies arise in the sound.

overdub A process whereby a track is recorded while listening to already recorded material on the same or a different tape recorder.

override See *ducking*.

overtone A harmonic frequency component of a complex tone at a frequency higher than the fundamental. This can involve both harmonic as well as nonharmonic overtones.

PAC Perceptual Audio Coding. Bit reduction system from Lucent: 96 kbps.

pad Circuit in the input module (for example, in a mixer) with the purpose of preattenuating the signal.

PAL Phase Alternation Line. European TV system.

pan-pot Potentiometer in the mixer for "moving" the signal between the left and right channel.

parallel port Digital input or output port with multiple lines, allowing multiple bits of data to be transferred simultaneously.

partial One of a group of frequencies, not necessarily harmonically related to the fundamental, which appear in a complex tone. Bells, xylophone blocks, and many other percussion instruments produce harmonically unrelated partials.

patch Connection of circuits via external connections.

PCM Pulse-code modulation. A form of modulation in which the information is described by the number and duration of impulses.

peak The maximum positive or negative dynamic excursion from zero of any time waveform. Sometimes referred to as "true peak" or "waveform peak."

peak-hold Circuit that for a shorter or longer period of time is in a position to maintain a peak value display on an instrument.

perceptual coding A principle for low bit rate coding based on the masking abilities of the ears.

periodic signal A signal is periodic if it repeats the same pattern over time. The spectrum of a periodic signal always contains a series of harmonics.

phantom supply Method for DC power supply of electrostatic microphones via the signal cable (normally 48 volt).

phase The different values that alternating current or alternating voltage runs through during a period.

phon Logarithmic unit for loudness.

phono plug/jack Plug and jack for one conductor and shielding.

pink noise Electrical noise signal for test purposes; contains constant energy per octave.

PIPU Punch In Punch Out.

pitch Subjectively perceived tone frequency.

PLL Phase-locked loop. A circuit that ensures a stable frequency in relation to a reference.

polarity Referring to the positive or negative direction of a signal. In all kinds of stereo and surround sound production it is important that microphones have the same polarity or else the imaging is totally blown.

post production The entire post-processing phase (including editing, sound effects, music, etc.) for video and TV productions.

pot See *potentiometer.*

potentiometer Adjustable resistance used, for example, in controlling levels.

PPM Peak program meter. A meter with a time constant of 5 or 10 ms.

precedence effect See *Haas effect.*

pre-emphasis See *emphasis.*

presence The frequency range around 2—5 kHz. Highlighting of this range causes voices and instruments to stand out in the acoustic image.

pressure zone As sound waves strike a solid surface, the particle velocity is zero at the surface and the pressure is high, thus creating a high-pressure layer near the surface.

psychoacoustics The study of the interaction of the human auditory system and acoustics.

punch-in A rerecording of a short sequence in an already recorded program.

punch-out See *punch-in.*

Q factor Measure for the slope of a resonance top.

quantization Digitalization of a signal. Conversion of the analog signal to numbers that express the values measured at the time of the samplings.

rack Cabinet or frame in which devices are installed.

RAID Redundant array of inexpensive disks. Method for simultaneous operation of multiple hard disk drives so that they work as a fault-tolerant unit.

RAM Random-access memory. Digital storage where data can be arbitrarily stored and retrieved.

RASTI Rapid Speech Transmission Index. See *STI.*

R-DAT Rotary-Head Digital Audio Tape recorder.

refraction The bending of a sound wave from its original path, either because it is passing from one medium to another with different velocities or caused by changes in the physical properties of the medium, for example, a temperature or wind gradient in the air.

resolution The number of bits per sample of a digitized signal.

RF Radio frequency. Name for frequencies in the range of 30 kHz to 3000 GHz. This is the range for electromagnetic waves applied to radio communication.

RMS Root mean square, the effective value of a signal.

roll off The attenuation of a high-pass or low-pass filter is called roll off. The term is mostly used for high-frequency attenuation.

room mode The normal modes of vibration of an enclosed space. See *mode.*

routing switch A switch for the routing of signals.

RT60 Reverberation time based on the 60 dB attenuation of the sound after the sound source has stopped.

RTP Real-time Transport Protocol. Protocol for the transfer of data for immediate use.

S/N See *signal-to-noise ratio..*

S/PDIF Sony/Philips Digital Interface.

sample rate converter (synchronous) Converts digital signals with the same reference from one sampling frequency to another.

sampling In digital technology, entry of the signal into discrete values in a table. A single value is called a sample.

sampling frequency The number of samples that are extracted per second. In professional sound equipment: 48 kHz. For CDs and the like: 44.1 kHz.

SCART plug A special 21-pin plug/jack on some video tape recorders and TV units.

SCSI Small Computer System Interface. Ultimately a purely computer-related standard, for the controlling of hard disk drives and such like. It can, however, also carry sound information.

SDDS Sony Dynamic Digital System. Digital sound system for movies with a total of 8 sound channels: L, CL, C, CR, R, SL, SR and subwoofer. SDDS is bit-reduced with the ATRAC system.

SECAM Systeme En Couleur Avec Memoire. Television system.

sensitivity Expresses the size of a signal that a device requires in order to reach nominal level.

separation Degree of signal-related separation between tracks, channels, etc.

sequencer A device on which information about sequences can be recorded, but not the sequence itself.

SFX Special effects.

shelving filter See *filter, shelving*.

short-form content Advertising, commercial, promotional, or public service related material or essence. Also termed "interstitial" content. The typical duration is less than approximately two to three minutes.

shuttle mode As related to tape transport or DAW track handling, the running of the recording back and forth between two points.

signal-to-noise ratio Ratio of the nominal signal level to the undesired noise level. Expressed in dB.

sinusoidal tone The most simple harmonic tone, consisting of only one frequency.

slap back A discrete reflection from a nearby surface.

slating Marking on a recording before a take.

SMPTE Society of Motion Picture and Television Engineers. International association.

snake For PA systems, the bus cable between scene microphones and the mixer in the concert hall.

SNG Satellite news gathering. Direct news transmission to main stations via satellites.

SOL Saturation output ;evel. The magnetization level at which a magnetic tape is saturated.

sound field microphone A single unit microphone with four cardioid condenser microphone capsules shaped like a tetrahedron. This configuration is also called the A-format. By processing the signals the B-format is formed equivalent to three figure-8 microphones and one omnidirectional microphone, all in a coincident position. A countless number of directional characteristics are obtained by combining the signals in different ways.

sound pressure level (SPL) The sound pressure level of a sound in decibels, equal to 20 times the logarithm to base 10 of the ratio of the RMS sound pressure to the reference sound pressure 20 µPa.

sound pressure A dynamic variation in atmospheric air pressure. Sound pressure is measured in pascals (Pa).

SP Special performance. Relates to improved versions of existing video formats.

spherical wave A sound wave in which the surfaces of constant phase are concentric spheres. A small (point) source radiating into an open space produces a free sound field of spherical waves.

splitbox A unit that distributes the signal from a microphone to multiple mixers or tape recorders.

square waves Waveform for acoustic or electrical signals that in terms of frequency contain a fundamental tone and all the odd-numbered harmonics.

stereo A technique intended to created a spatial sound impression by the use of two or more channels.

STI Speech Transmission Index. Special measurement method and measurement equipment for the objective determination of speech intelligibility in a room. In a good room, the STI value ought to be greater than 0.6 (on a scale of $0 - 1$). Practical (reduced) versions that can be implemented in handheld measurement devices are RaSTI (Rapid Speech Transmission Index) and STIPA (Speech Transmission Index, PA systems).

STIPA See *STI*.

streaming audio Sound that can be played back immediately in (almost) real time from the Internet.

subharmonic Subharmonics are synchronous components in a spectrum that are multiples of 1/2, 1/3, or 1/4 of the frequency of the primary fundamental.

supersonic See *ultrasound*.

surround sound Designation for certain multichannel systems for film and TV stereo sound.

sweet spot The optimum listening position in front of the stereo- or surround speakers.

symmetry attenuation The attenuation of induced electrical noise signals that is attained in a balanced connection.

synchronizer A unit that by the use of a time code can deliver control signals for the time-related locking of, for example, tape units.

tape speed The speed at which a magnetic tape passes the magnetic heads in a sound or video tape recorder. Specified in cm/s or inch/sec.

tape width The width of a magnetic tape.

taping Outlet from mixer channel, for example aux-send. The expression is used by radio broadcasters.

target loudness A specified value for the anchor element (in some systems dialog level), established to facilitate content exchange from a supplier to an operator.

TB Talk back. A system via which the control room can give oral messages to the studio.

TC See *time code*.

TDIF Tascam Digital Interface. Factory-initiated interface, 8-channel parallel, unbalanced in a D-sub connector, cable max. 15 m [45 ft].

telephone hybrid Unit that adapts a telephone to a mixer console.

THD Total harmonic distortion.

THD+N Total harmonic distortion plus noise. An electro-acoustic measure of (unwanted) signals from distortion (occurring, for instance, by the clipping of the signal) and the noise in the given audio channel. Specified in dB below the main signal or in a percentage.

third-octave band A frequency band whose cutoff frequencies have a ratio of $2^{1/3}$, which is approximately 1.26. The cutoff frequencies of 891 Hz and 1112 Hz define the 1000 Hz third-octave band in common use.

third-octave filter See *filter, third-octave*.

tie-line Connection line for signals (for example, between editing rooms).

TIM Transient intermodulation distortion; occurs when the feedback circuitry is not acting fast enough when the amplifier is exposed to fast transient signals.

timbre The quality of a sound related to its harmonic structure.

time code Digital code that marks time with information on frames, seconds, minutes, hours, etc. Examples: SMPTE, EBU, MTC.

time constants In transmission and recording, corner frequencies of de-emphasis are often defined by time constants, for example 3180 μs. The frequency is 50 Hz, since the time constant $\tau = 1/2\pi f$.

tinnitus Ringing in the ear or noise sensed in the head typically caused by excessive exposure to high sound levels.

tolerance The maximum permissible deviation from the specified quantity.

TP Abbreviation for true peak (unweighted). The term is used in digital audio metering and is then related to full scale. See *true peak*.

transfer function The output to input relationship of a structure.

transformer An electronic device consisting of two or more coils used to couple one circuit to another.

transient A relatively high amplitude, suddenly decaying, peak signal level.

true peak The maximum absolute level of the signal waveform in the continuous time domain, measured per ITU-R BS.1770. Its units are dB TP meaning decibels relative to nominal 100%, true peak.

truncation Cutting off of superfluous bits without weighing their value.

twisted-pair A way to ensure balancing in cables.

ultrasound Sound at frequencies above the audible range, that is, above about 20 kHz.

unbalanced A connection where the chassis/ground is a part of the circuit between electro-acoustic devices. Also referred to as single ended.

varispeed Variable speed, generally in playback systems.

VCA Voltage-controlled amplifier.

VCR Video cassette recorder.

voiceover See *ducking*.

volt, V Unit for electrical voltage.

Vpp Measure of the magnitude of an electrical voltage from peak to peak.

VTR Video tape recorder.

VU Volume Unit, a level meter (standard volume indicator) for audio recording; time constant: 300 ms.

watt, W The unit of electrical acoustic power.

wav File format for sound.

waveform The waveform is the shape of a time domain signal as seen on a DAW screen or an oscilloscope screen. It is a visual representation or graph of the instantaneous value of the signal plotted against time.

weber, Wb Measurement unit for magnetic flux.

weighting A measurement is weighted when the signal has passed through a filter (for example, an A filter).

white noise A signal that contains all frequencies, with constant energy per Hz.

woofer Bass loudspeaker.

working level That signal level at which a circuit is nominally modulated. Also called the "0-level."

WRMS Weighted RMS value. The abbreviation is also used for the RMS value of the power.

XLR Professional cable connector for audio. Pin 1: shield, pin 2: +, pin 3: −.

Zwicker Loudness A technique developed by Dr. E. Zwicker for calculating a real-time estimate for the loudness of sound as perceived by the human ear. This methodology is primarily used in connection with calculation and perception of noise exposure.

Index